INVISIBLE ALLIES

Bernard Dixon
INVISIBLE ALLIES

Microbes and Man's Future

TEMPLE SMITH • LONDON

First published in Great Britain 1976
by Maurice Temple Smith Ltd
37 Great Russell Street, London WC1

© 1976 Bernard Dixon

ISBN 0 85117 0900

Printed in Great Britain by offset lithography by
Billing & Sons Ltd, Guildford, London and Worcester

Contents

Preface

We face global problems on a gigantic scale — problems of energy, resources, population, food and environmental destruction. We are tempted to imagine that only initiatives on the grand scale — big technology, summit meetings, world plans — are commensurate with the difficulties confronting us for the foreseeable future. Certainly, radical political and technological action is needed. But it also behoves us to turn our eyes from the very big to the very small, and to consider the many ways in which the tiniest of all living creatures, the microbes, can assist us in our plight. We are already utterly dependent on micro-organisms for our health, wellbeing and even our very existence. Intelligently harnessed, the same organisms and countless others of unique versatility and exotic skill could make a massive contribution to solving our problems today. We are just beginning to appreciate this vast potential. Yet we still tend to believe, by ungenerous reflex, that microbes are utterly malevolent; we seek to devastate them by ruthless cleansing wherever possible, and have now come near to driving some of them to extinction. That scandalous paradox is what this book is all about.

I want to thank Katherine Adams, who organized my thoughts; Mike Bessie, Tony Loftas, Martin Sherwood, Peter Stubbs, Maurice Temple Smith and Nick Valery, who read and made many helpful comments on parts of the book; and the following copyright owners for permission to quote passages from the publications named:

René Dubos (*Mirage of Health*), Cambridge University Press (Macfarlane Burnet's *Natural History of Infectious Disease*), Faber & Faber (C.L. Duddington's *Microbes as Allies*), Odham's (David Lloyd George's *War Memoirs*), and Little, Brown and Co (Hans Zinsser's *Rats, Lice and History*). I am particularly indebted to General Duckworth and Alfred Knopf for permission to use 'The Microbe' from Hilaire Belloc's *More Beasts for Worse Children*. Finally, my grateful thanks to the editors of *The Spectator* and *World Medicine* for allowing me to draw on my own material previously published in those journals.

Billion is used throughout this book to mean one thousand million. Dollars are US dollars unless otherwise stated.

1 Ubiquitous Benevolence

The microbes have always had a bad press. From the moment they first began to be identified just over a century ago, it seemed obvious that infinitesimal forms of life, so small that they could be seen only through a microscope, must be up to no good. They were soon incriminated as agents of disease, death and decay, their malevolent activities rumbled at last, after centuries of miasmata and mystery, by the dedicated labours of Louis Pasteur, Robert Koch and the other great 'microbe-hunters'. One by one, the microbes responsible for man's historic plagues—tuberculosis, diphtheria, cholera—were isolated and given pathological identity tags. Next, vaccines and drugs were devised to combat the dastardly pathogens. Together with improved hygiene and sanitation, this work led to the conquest of infectious disease, one of the boldest chapters in the history of applied science. The clarion call was: exterminate the microbe, and the scientists most honoured in this field were those who were outstandingly successful in doing so.

This is still the mood today. Watching television advertisements, or thumbing through the glossies, one is struck by the phrenetic message, glamorously purveyed, that one of the most responsible things you can do for yourself or your family is to vanquish microbes from your body and home. Not only is the best disinfectant that which kills 'all known germs': there is also the insistent advice to spray or dowse every crevice of your body in cleansing liquids ('approved by science') that will seek out and destroy the last vestiges of microbial life. Products have even been developed to do battle with the 'germs', which, advertisers try to persuade us, are proliferating clandestinely inside our domestic central heating systems.

The anti-microbial gospel has astonishing potency.

Part of the explanation must be a deep-seated desire for
scrupulous cleanliness, coupled with dread of the unseen,
a belief that dirt so tiny as to be invisible to the naked eye
must be much filthier than that which we can see. The
message—a dangerously wrong-headed one—has arisen
through chance historical circumstances. For the fact is
that most of the microbes that surround us from birth until
after death (however vigorous our offensive measures
against them) are vastly more beneficial than harmful,
more constructive than destructive. Microbes were the
earliest forms of life on earth, and they are crucial in
sustaining our existence today. Yet, while we proudly
remember Pasteur, Alexander Fleming and their fellow
microbe hunters, the names of Martinus Beijerinck and
Serge Winogradsky, who first demonstrated our total
dependence on the microbial world, have been all but
forgotten.

What is especially intriguing about this imbalance is
that Pasteur—a chemist, not a doctor—did his initial
microbiological work on the action of micro-organisms
in souring milk, cheese-making and brewing. Only later
did Pasteur succeed in translating his central discovery
from these investigations—that specific types of microbe
were associated with particular chemical changes—into a
similar understanding of the basis of infectious disease.
His most fundamental revelation was of the prime importance of microbes in furthering organic processes,
rather than in fomenting illness.

But while it was Pasteur the scientist who ended speculation and miserable helplessness by revealing the real
cause of infectious disease, it was the medics who carried
through much of the subsequent work in the young discipline of microbiology. They were responsible for track-
ing down one 'germ' after another and then devising
vaccines and drugs to attack the invisible enemy. It was
doctors too who wrote most of the books about the newly
emerging science. Small wonder, then, that they concentrated on the glorious march of medicine in purging the

world of microbes, and ignored the other side of the picture. Even William Bulloch's classic, *The History of Bacteriology*—first published in 1938, reprinted in 1960, and still the only historical account of its sort—ignores almost entirely the overwhelming evidence of the positive value of microbes, while comprehensively documenting their demerits.

The millions of bacteria that live on the human skin illustrate very clearly our warped judgements. However much soap and water we use, whatever the deodorants and anti-perspirants we apply to our surfaces, our bodies are heaving with microbes—some 100,000 of them to every square centimetre. A recent survey showed that there are more bacteria on and in every person's body than there are human beings on earth. Clearly, this teeming microbial life cannot be harmful. On the contrary, it is essential to health. Consider, for example, the special situation on the buttocks. Once, or sometimes twice or more times a day, millions of the bacteria that live in our intestines are deposited around the anus during defaecation. Within two hours, none remain: they have all been obliterated by the bacteria that occupy the skin as their home territory. Skin bacteria are essential scavengers. And if we repeatedly try to eradicate them, by deploying hexachlorophane deodorants enthusiastically into armpit and crotch, the result may well be to kill of the normal residents and allow less favourable types to migrate to the region, causing discomfort, itching and eczema.

Versatility and industry in the soil
Every saltspoonful of soil on this planet contains at least two billion microbes. That is an incomparably larger population than those of the few microbial species we believe to be nuisances. Just as the astronomical hordes of bacteria covering our persons have no ill effects, and indeed are useful to us, so the microbial community in the soil is exquisitely well adjusted for coexistence with the plants, and animals, that it supports.

One of the key functions of soil micro-organisms is to break down dead plant and animal tissues and return their nutrient chemicals to the soil. Yet in high summer a field of, say, turnips or potatoes can grow vigorously in intimate contact with billions of creatures specifically capable of destroying these vegetables. Only if and when the tissue has died can the scavenger organisms operate, and then they do so swiftly and efficiently. As with our skin inhabitants, there is a disturbing lesson here for those who believe in shattering the natural balance of the microbial world by chemical overkill.

Though recognized only during the past century, the scavenging role of microbes has been built up over millions of years of evolution. One aspect of this unique capacity is peculiarly relevant to modern, technocratic society. Innumerable new chemical compounds come into industrial and domestic use every week—drugs, solvents, food additives such as preservatives, colouring agents and flavours, pesticides, adhesives, paints and many more. Some of these soon find their way into the soil. Many others turn up there eventually or pass into the oceans. They have done their job, and we believe we can forget about them, as they are washed away into an infinite sink. The truth is, of course, that the world is finite. It is therefore desperately foolish to believe that the environment is a limitless sewer, into which we can jettison an endless succession of novel chemicals. The extent to which we have been able to do so with impunity in the past is a tribute to the incredible abilities of microbes to decompose and detoxify all manner of chemicals, even those that have never previously existed in the natural world and are the latest products of innovation by the chemical industry (Chapter 7).

What actually happens when insecticide XB44, last month's best buy at the garden centre, leaches into the garden soil from the apple trees where we put it, is that out of the teeming micropopulation a bacterium or fungus emerges which can break down this particular chemical.

The scavenger proliferates temporarily, destroys XB44, and then resumes its former position in the balanced population. Such minor shifts in the diverse working community of the soil go on continually, as the scavengers deal with both natural detritus and the increasing variety of man's technological debris. Similar processes occur in the rivers, lakes and oceans. In soil, the picture is one of a huge population in which innumerable microbial species perform specialized, interdependent tasks of disposal. Vested in this living pattern is both the capacity to maintain soil fertility, despite the physical and chemical extremes that can occur even in normal times, and the versatility to deal with most of the chemical insults that we, from time to time, choose to apply to field and garden.

Microbes and global crisis

A hostile attitude towards microbes, it seems, is both ill-founded and short-sighted. The need for reassessment is urgent, because of the considerably greater contributions that the microbial world, fully and intelligently harnessed, could make in the decades ahead to the solution of our most pressing global problems: energy and resource crises, food shortage and pollution. Each of these problems demands that we acknowledge for the first time the disproportionate beneficence of the very smallest living creatures. Certainly, none of them can be solved if we fail to make this readjustment in our thinking.

First, the microbes' scavenging role is already vastly important, albeit largely unrecognized. But such is the metabolic potential of the sub-microscopic world that the skills of its creatures could be channelled much more extensively and consciously to help us deal with the excreta, natural and technological, of modern society (see Chapter 7). There are those who argue that massive pollution is an inevitable consequence of the size of modern communities and the centralization of industry, and that by returning to small-scale living, with self-sufficient technology, the environmental rape characteristic of city living can be avoided. That may well be true. But most people today do not wish to live in communes or pastoral communities. We have become too appreciative of the merits of our present system — whether those of science-based medicine, electricity supply or even food technology — to relish the prospect of exchanging them for days spent attending to our individual solar roofs and sewage farms.

True, a healthy movement towards what used to be called 'alternative' ideas is already apparent. But sudden and drastic changes in the pattern of living seem unlikely in the foreseeable future. Thus pollution, despite the progress of recent years, will remain a considerable problem. Why, then, do we not harness microbes to a far greater extent than at present to break down the toxic and waste materials that we now allow to become offensive pollutants?

Why not go further and turn our effluents and wastes into food? Already, there are signs of a vigorous growth of interest in the possibilities of cultivating yeasts and algae on a huge scale as a source of food—food that is frequently rich in protein and vitamins. Many microbes can be raised on unwanted materials that we usually throw away or discharge as environmental pollutants. Yeasts, for example, can be reared in the 'sulphite liquor' that is a bulk waste product of paper manufacture, while algae may be cultured in sewage. If we choose algae, there is the additional opportunity of tapping with greater efficiency than usual the rays of the sun—a virtually inexhaustible source of energy, which we largely waste.

One exciting prospect, which links our global hunger for both food and energy, is to produce methane gas from sewage. Each of us contributes about ninety grams of organic matter every day to the public sewage system. Using the 'heated digestion' method of processing, this can yield some thirty litres of methane (even when the breakdown process is only fifty per cent efficient) which may then be burned and used to generate electricity. Moreover, the material left behind is liquid compost, an ideal fertilizer. About two-thirds of Britain's sewage treatment plants have adopted this technique, which provides more than enough methane to run the pumps and other machinery. But many countries do not use the process to anything like the same extent—and even in Britain billions of kilos of sewage are simply dumped into the rivers and sea each year. Such a resource, valuable and gratis, should not be so shamefully wasted.

We also face tremendous opportunities to improve the contributions of the microbes to conventional food-raising. Traditional agriculture depends entirely upon the recycling of the chemical elements—principally nitrogen, sulphur, carbon and oxygen—by bacteria and other microbes. Above all, it rests upon the 'fixing' of nitrogen from the air into substances that plants can absorb and use for growth. Some of the microbes responsible occur in soil and water. Others live in nodules on leguminous plants such as clover, in the mutually supportive relationship known as symbiosis. Recent developments in understanding how they fix nitrogen, in inoculating nitrogen fixers into soils lacking them, and potentially in creating novel plants able to fix nitrogen, have heightened the possibility of boosting greatly this aspect of microbial munificence.

The world shortage of chemical fertilizers that has arisen in recent years makes such work all the more urgent. We have to reckon with the inconvenient fact that nitrogen fertilizer is in part a petrochemical product. The oil industry provides the starting materials (mainly methane and naphtha) that yield the hydrogen used to synthesize ammonia and nitric acid from which, in turn, nitrogen fertilizers are manufactured. Hydrogen can also be extracted from sea water, by electrolysis, but this process is extremely costly—particularly in energy—in relation to its returns.

It was the action of the oil sheikhs in the autumn of 1973 that brought home to the affluent West the enormity of the energy crisis facing the world. The decision to exploit

their control of petroleum for political and financial
purposes was new: the underlying situation was not. We
had allowed ourselves to base modern industry and thus
society to a perilous degree upon a single natural resource, which, by accidents of history and geology,
happened to occur in profusion in what was formerly
considered a politically inferior region of the world. But
that commodity existed in finite quantities and would
thus inevitably rise in cost, probably very steeply, amid
the political realism of the modern world.

There is a nice irony here. Petroleum is thought to have
been created originally by bacteria, transforming sedi-
ments in the oceans millions of years ago. Today, it is
being prospected for by lowering cultures of other bacteria, responsive to vanishingly small concentrations of
petroleum vapours, into suspect strata. In times of scarcity,
microbes are thus proving valuable as ultra-sensitive
detectors of the petroleum which was produced by microbes
in the first place.

Pressures of adversity

It is in moments of crisis that we re-examine the material
premises and conventional wisdom behind our present
ways of doing things — and sometimes come up with start-
ling, yet startlingly obvious, conclusions. So it was, in the
aftermath of the Western world's initial oil crisis, in
February 1974 that Dr William Smyers wrote to the American journal *Science* to urge a peculiarly persuasive use
of microbes in the face of international adversity.

'Adding alcohol to gasoline', Dr Smyers wrote, 'would
help to make the United States self-sufficient in energy
and would help build up an industry which could be ex-
panded hurriedly in case of emergency.' Ordinary (ethyl)
alcohol has an octane number of 99. Even a mixture of
equal parts of alcohol and 70-octane petrol has an octane
number as high as 88. In prewar days many European
countries passed laws requiring the blending of alcohol
with motor fuel, and the consumption of 'power alcohol'

throughout Europe in 1937 amounted to more than 500 million kilos.

Decades of cheap oil from the Middle East meant, however, that such stratagems were unnecessary. There are many other, apparently more important, uses for alcohol than pouring it into one's petrol tank. Besides, internal combustion engines would have to be modified to burn alcohol most efficiently as a fuel. Yet, triggered by the oil sheikhs into realizing our precarious dependence on oil, what was formerly a theoretical irrelevance began to appear both realistic and attractive. Alcohol is one of the most common products of yeasts and bacteria, and only the pressures of adversity have been needed to make us see the wisdom of developing this process on a grand scale.

The analogy with penicillin manufacture during wartime is irresistible. The *Penicillium* mould studied by Fleming in the late 1920s yielded impractically tiny amounts of penicillin. But once the medical value of penicillin, and thus the need to boost productivity, became clear, microbiologists were able to raise yields enormously. During the early 1940s, by searching for more generous strains of *Penicillium*, and by growing them in improved culture media, they increased the output of the life-saving antibiotic many thousandfold. Similar examples of innovation stimulated by the exigencies of war include food yeast production at a factory opened by the UK Colonial Office in Jamaica in 1944, and the microbial manufacture of glycerine in Germany during the First World War.

Oil is not the only commodity of whose terrestrial scarcity we have become aware recently. Depletion of several vital minerals looms, too, particularly of metals such as copper and uranium. That is why mining engineers throughout the world are re-examining an intriguing series of methods which exploit bacteria to leach out metals from low-grade ores, inaccessible ores and waste materials that are usually discarded during mining. Conventional

recovery techniques have been devised to process high-
grade ores. But in the world as it now is, and will be, microbial processes look increasingly attractive. Though slower
than orthodox methods, they are extremely cheap in both
capital cost and power requirements.

Sulphur is one element whose existence we owe largely
to bacteria. Some of the world's deposits are volcanic, but
the major ones are associated with former seas that dried up
during geological history. The Texas deposits, which
today yield some ninety per cent of the world's sulphur,
were created at a time when part of the Gulf of Mexico was
being desiccated. During this bizarre period, bacteria
converted calcium sulphate (the substance now known in
different forms as gypsum and plaster of Paris) to sulphur.
Today similar transformations appear to be responsible
for the production of sulphur in Libya, where the Arabs
harvest it from several lakes. Prospects for producing
sulphur industrially by microbiological methods are now
being energetically investigated.

Does extinction matter?

One of the best known books about microbes is Hans Zinsser's *Rats, Lice and History,* first published in 1935. A spirit-
ed account of typhus fever and other infections, it portrays
microbes in the context of 'the struggle for existence between different forms of life. Man sees it from his own
prejudiced point of view; but clams, oysters, insects, fish,
flowers, tobacco, potatoes, tomatoes, fruit, shrubs, trees,
have their own varieties of smallpox, measles, cancer, or
tuberculosis. Incessantly, the pitiless war goes on, without
quarter or armistice – a nationalism of species against
species.'

Zinsser concedes here the anthropocentricity of our
reflex attitude towards the microbes. But he does not go
so far as even to acknowledge the symbiosis, cooperation
and division of labour that occurs throughout the microbial kingdom, or our utter dependence upon it. Zinsser
would doubtless be horrified to learn of our modern plight,

in which we begin to see microbes as potential saviours which could help us avert catastrophe. There is little doubt about his likely answer to another, perplexing problem with which we are likely to be confronted very shortly. The dilemma is simply whether or not to render extinct, by calculated choice, a particular species of life: the smallpox virus. So successful has the World Health Organization's programme against smallpox been in recent years that the possibility of total eradication of the disease is imminent. We tend to think of smallpox as no more than a peculiarly vile disease, a deadly plague to be abolished as far as possible. But if and when we do eradicate this (or any other) infectious disease completely, we shall also render extinct the microbe that causes it.

Whether to let this happen is a much more profound dilemma. True, many other species of living creature have become extinct, many by accident, some by our neglect. At the present time twenty thousand plant species — about ten per cent of the earth's flora — stand in danger of extinction. The extent to which any of those plants are allowed to disappear, for ever, from the face of the earth will depend on our lack of interest, our ignorance and conversely our energy in taking action to protect threatened species. But to drive a particular organism into ever smaller areas by conscious campaigning, and then to make the final decision to abolish the last remaining representatives of a species (even a pathogenic one) which has existed at least since the Middle Ages, is a very different matter. It could well be that, whatever its ravages in the past, we may in years ahead come to need the smallpox virus to help us *fight* disease (Chapter 11). Despite being one of the tiniest life forms, such a virus is incredibly complex and could never again be recreated artificially in the laboratory. Just as we seek to conserve rare plants and animals, therefore, there is a strong case for acting — against all common sense perhaps — to preserve the agent of smallpox.

But such is our antagonism towards the microbial world

that common sense, rather than prudent judgement, may win the day. It is indeed amazing, despite our *absolute* reliance on microbes biologically and economically, and despite our daily exploitation of them for comfort and profit, that their only well-publicized exploits are their occasional attacks on *Homo sapiens*. The most trivial influenza epidemics — even those that fail to develop — receive banner headlines in the press. Outbreaks of genuinely serious disease are exaggerated in their importance, so evil are the guilty micro-villains thought to be (p. 213). And the fashionable ploy is to put all manner of bodily afflictions, from sunstroke to a hangover, down to that ever-ready scapegoat, 'a virus'.

This book is an attempt to redress the balance of adverse publicity surrounding the microbes. I shall not try to minimize the predations of some of their species: microorganisms have swung the course of history through such scourges as dysentery and typhus. What I *shall* argue is that man's biological dependence on microbes eclipses their unfriendly acts, real and imagined; that most of these apparent hostilities are our fault anyway; and that, rather than seek on every occasion to destroy microbes by indiscriminate slaughter, we should strive to live in harmony on earth with our invisible cohabitants. Above all, we should become much more — and much more actively — conscious of the skills and versatility the microbes offer us as collaborators in resolving our unprecedented global problems.

2 Small but not Simple

The Microbe is so very small
You cannot make him out at all,
But many sanguine people hope
To see him through a microscope.
His jointed tongue that lies beneath
A hundred curious rows of teeth;
His seven tufted tails with lots
Of lovely pink and purple spots,
On each of which a pattern stands,
Composed of forty separate bands;
His eyebrows of a tender green;
All of these have never yet been seen —
But Scientists, who ought to know,
Assure us that they must be so. . . .
Oh! let us never, never doubt
What nobody is sure about.

Hilaire Belloc

The development of science is an untidy, irregular affair. Its propellants range from individual researchers' brilliant intuition and fortuitous mistakes, to the external forces of social, political and economic need. One factor often forgotten is the practical breakthrough which suddenly facilitates investigations that were formerly impossible. The invention of the microscope, and its subsequent improvement to provide increasing observational power and resolution, is an outstanding example. At a stroke, the introduction of this piece of optical equipment permitted inquisitive individuals to begin peering for the first time at the fine structure of their bodies and the tissues of plants and animals, and thus to discern the exquisitely fashioned cells upon which all life is built. Then, towards the end of the seventeenth century, the microscope revealed the

21

hitherto unseen world of 'animalcules', as their discoverer
Antony van Leeuwenhoek called them. Visible only
through the microscope, these ubiquitous inhabitants of
our bodies and our environment became known, by
definition, as microbes.

Van Leeuwenhoek was a linen merchant in Delft,
Holland. Though not a trained scientist, he turned to the
unusual hobby of grinding and mounting unusually
powerful lenses, which he used to study all manner of
materials — cow dung, scrapings from his teeth, seminal
fluid, infusions of pepper, plant leaves and countless
more. Though compound microscopes with more than
one lens, similar to today's models, had been built previously, van Leeuwenhoek's designs were simple instru-
ments. They were little more than skilfully mounted single
lenses — superior magnifying glasses.

Yet they allowed him to observe, and fall in love with,
the busy world of animalcules.

Microbiologists still argue over precisely what van
Leeuwenhoek did see through his microscopes (constructional details of which he always kept secret). Some of his
work is indeed difficult to fathom — whether, for instance,
he anticipated the modern technique of 'dark ground'
microscopy — but we do have on record the letters and
drawings describing his investigations which he sent to
the Royal Society in London. From these, it is apparent
that the enthusiastic Dutchman was the first person in
history to observe the creatures we now know as bacteria.

It was nearly two hundred years later, however, with
the work of Louis Pasteur, that the era of modern bacteriology really began. As we have seen, Pasteur took a crucial step in establishing that particular microbes conducted
particular organic transformations: bacteria souring
milk, and yeasts producing alcohol by fermentation. Later,
he and Robert Koch and others uncovered many further
associations between species of bacteria and corresponding
infections.

Pasteur's other historic achievement was to disprove

the venerable theory of spontaneous generation. Two centuries earlier the Italian physician Francesco Redi had taken the first step, showing that meat protected from egg-laying flies with fine gauze did not develop maggots — which had previously been thought to arise spontaneously in the putrid meat. Pasteur went one stage further. In a series of dramatic public demonstrations, which he often used and greatly relished, he proved that broth that had been sterilized by heating (but which would otherwise have supported living bacteria) remained free of life as long as microbes were prevented from entering. Some of Pasteur's sealed flasks of broth, which were sterile in 1864, have been preserved and are still sterile today.

The microbe is so very small...
Bacteria are intermediate in size among other microbes we shall be considering (viruses at the bottom end of the size range, and algae and fungi at the other). They are, nevertheless, so minute that a box measuring one cubic inch could contain nine trillion (9,000,000,000,000) of a medium-sized variety. Special units had to be found to measure and describe them when they were first discovered: the originators of the metric system had not envisaged the need for such divisions. The unit is known as the micrometre (usually written μ), which is a millionth of a metre — or a thousandth of a millimetre.

The smallest of the common bacteria, spherical types called *cocci*, are about 2μ in diameter. The rod-shaped *bacilli* can be up to 10μ long. A few bacteria are filamentous, and some filaments measure as much as 500μ in length. At the lower limit of size, certain species of mycoplasmas are only 0.135μ in diameter.

Do not be misled by these figures. There has long been a tendency to think of bacteria and their microbial colleagues, by virtue of their tiny dimensions, as very, very simple forms of life. Textbooks speak of 'lower organisms' and 'primitive metabolism'. This gross unfairness can be traced to the dominance of medical people in (what has

always had to struggle to become) the science of microbiology. Medical bacteriologists seldom or never examine
the living creatures they cheerfully categorise as primitive
to check the foundations of their belief. What they actually
do, busy in the pathology lab, is to take samples of blood,
urine and similar materials, apply them to microscope
slides, pass the slides through a flame to 'fix' the smears,
and then dowse what remains in huge concentrations of
various dyes. The dead and frizzled remnants of some
pathogens take up certain stains, thereby revealing their
presence. That is all the medical bacteriologist wants to
know. He has no interest beyond confirming that his
patient's sputum does or does not contain tubercle bacilli.

But the distorted vestiges of bacteria prepared in this
way, with all cellular detail obliterated by a blanket of
purple or red dye, bear no resemblance whatever to the
same creatures when healthy and living. Medical students'
experience of bacteria is limited to spending a few hours
preparing such distorted vestiges of the few species that
happen to cause disease or inconvenience to man. Small
wonder, then, that they acquire a similarly imbalanced
view of the microbial world.

For the fact is that micro-organisms can convincingly
claim to be among the most successful representatives of
life on earth. They have already existed for much longer
than man himself — some billion years, as against a million
or so. They have colonized an incredible range of habitats,
surviving and adapting to physical and chemical conditions that man could not possibly withstand. And they
manage to pack all of the essential functions of life, from
reproductive to excretory capacity, into a single cell. We
and other 'higher' forms of life need many specialized
tissues and organs for these purposes. Microbes are not
simple but economical. They are not primitive but decep-
tively sophisticated, not lower organisms but highly
efficient ones.

Consider the abundant merit of being tiny. The volume
of an object increases more rapidly than its surface area.

Conversely, the smaller an object becomes, the greater, proportionally, is its surface area compared with its volume. If we choose our units so that man has a surface area-to-volume ratio of twenty a typical bacterium can have a ratio of nine million. Because the foods necessary for a cell to grow, and conversely its waste products, must pass through the membrane surrounding the cell, there is clearly a substantial advantage in smallness. The many chemical processes of life can take place more rapidly in a midget than in a larger cell. This helps to explain the tremendous rate of the vital processes (metabolism) of bacteria, yeasts and other microbes, and has clear implications for man in harnessing their skills as synthetic chemists (Chapter 8) and disposers of our debris (Chapter 7).

Another key characteristic of micro-organisms is the stupendous rate at which they multiply. Many bacteria — such as *Escherichia coli,* one of the commonest and best known of those living in the human gut — can reproduce by dividing every twenty minutes or so. This means that one cell of *E. coli,* added to a thimble-full of broth, can increase a thousandfold in just over three hours. In about a day it can spawn billions of progeny.

Clearly, as a tiny cell grows in volume, there comes a point where the relatively declining surface area cannot accommodate a sufficiently rapid interchange of food and wastes. The initial, favourable relationship between surface and contents must therefore be renewed. Bacteria accomplish this by the cell dividing into two smaller daughter cells. Yeasts, on the other hand, multiply by budding. Again, the unique pace at which microbes reproduce is directly relevant to our interest in harnessing their abilities. A vast quantity of cellular material can be produced very quickly during microbial food manufacture, for example.

What microbes look like

One of the most striking features of the structure of bacteria, yeasts and many other microbes is the resilient cell

wall surrounding each cell. Though tough, the wall contains pores and is relatively permeable to the various substances that have to pass in and out of the cell. What actually
regulates this interchange is the much more delicate cell
membrane, lying within. The function of the wall is to
protect the membrane.

The success of this arrangement can be judged from the
force that must be applied to smash open microbial cells to
study their contents in the laboratory. Two of the methods
that have to be used are violent shaking with minute glass
beads or sand, and disruption by exposure to ultrasonic
vibrations. Another technique exploits the fragility of the
membrane when it is no longer protected by the cell wall.
Under these circumstances, the membrane will burst (by
osmosis) if the denuded cell is placed in distilled water,
which does not contain the same amount of dissolved
material as does the protoplasm inside the cell. Penicillin
acts against some bacteria by inhibiting production of the
cell wall, and can thus be used to prepare vulnerable cells
(protoplasts) devoid of their protective shield. The only
bacteria that exist normally without a cell wall are the
ultra-small mycoplasmas, which live in protected situations in the animal body.

Cell membranes are particularly important structures.
As well as regulating the flow of essential materials into
and out of bacteria, they play an active role in blocking the
transport of molecules whose movement, in either direction, is not in the interests of the cell. They also contain
substances involved in respiration.

The protoplasm inside a growing microbial cell is an
inferno of chemical activity. This centres around transla-
tion of the coded messages in the hereditary material into
the manufacture of new cell structures and the renewal
of the old. In the most intensively studied bacterium,
E. coli, the genetic material is deoxyribonucleic acid (DNA),
which occurs in the form of a long, circular fibril. As in
man, cabbages, leopards and all other creatures yet investigated, the cellular processes are all mediated by

enzymes, proteins that are manufactured in particles called ribosomes. Another nucleic acid, ribonucleic acid (RNA), plays the central role here, translating the hereditary instructions of the DNA into the assembly of different enzymes and other proteins.

Convention has it that a crucial difference between the 'primitive' bacteria we have been considering and such marginally higher microbes as yeasts, blue-green algae, fungi and protozoa (and indeed man himself) is that the latter have a membrane around a well-defined nucleus carrying the DNA. This arrangement is accompanied by rather more visibly organized manoeuvres by which the genetic material is passed to daughter cells when cells divide. It is deemed slightly superior because of its resemblance to nuclear arrangements in animal cells. A more convincing interpretation is that the really significant feature — that which not only assists some microbes in seeking out favourable environments, but which also helped the earliest cells to begin to propel themselves out of the primeval scum and colonize the land — is the flagellum, a long, thin, whip-like appendage that confers the skill of movement.

According to Dr Kenneth Bisset, it was the flagellum that unlocked the limitless potential of biological evolution:

> What an invention! At one blow, we have the potential of active motility, which separates the truly animate from the all-but-chemically inanimate; the basis of most sense organs, whereby higher animals appreciate their environment; and the essential component of the ciliated feeding mechanisms that permitted primitive animals of every group to gather the microscopic harvest of the seas, and obtain their food already ninetenths prepared, so as to reserve some of their metabolic potential for development.

Bacterial movement is a curious phenomenon. The fastest rate that has ever been recorded for a bacterium is equivalent to about 0.0001 mile an hour for you or me. That doesn't seem too impressive, even when compared with the speed of a rowing boat. But water is a viscous medium, and the effects of viscosity are considerably greater on small objects than on large ones. Propelling a bacterial cell through water is roughly analogous with rowing a boat across a sea of thick molasses syrup. It is amazing that bacteria manage to move at all.

Many of the other structures of the bacterial cell are geared to survival. Thus some bacteria carry pilli, similar to, but shorter than, flagellae. They are not for locomotion, but help the organisms adhere to surfaces. Some pilli are also involved in sex (p. 30). Other bacteria produce, outside the cell wall, thick capsules to protect themselves against adversity. Some have holdfasts (like those of seaweeds), by which they stick to surfaces that offer favourable conditions for growth. Then there are those structures specifically helpful in sustaining the opportunistic mode of life of bacteria — the art of multiplying explosively when food is available and then lying dormant during lean times. These include the extremely tough spores that allow bacteria to resist all manner of environmental hostility, from fierce heat to disinfectants. Spores also provide for long-term survival. Soil samples attached to pressed plants kept at Kew Gardens have been found to contain living bacterial spores dating back to the seventeenth century. Many bacterial cells contain special granules in which they can store energy or stockpile building blocks for new cellular material.

Those microbes termed the most primitive, are, it seems, rather more complex than they are often represented. Turning to the types normally considered somewhat advanced, we find an increase in both stature and specialization. The blue-green algae, for example, are larger than bacteria and contain pigments such as chlorophyll in chloroplasts that harness the sun's rays in the process of photosynthesis, as in green plants. Much more complex are such creatures as the euglenoids, claimed by botanists as algae and by zoologists as animals. The best known example, *Euglena gracilis*, has considerable internal differentiation, with a mouth-like opening and gullet, chloroplasts and a light-sensitive eye-spot that allows the creature to orient itself towards the light. Then there are the relatively elegant and sophisticated protozoa, such as amoeba, with its false feet (pseudopodia) for movement, food vacuoles, and contractile vacuoles. They are firmly categorized as single-celled animals. Those found in profusion in soil and pond water range from 5μ to 50μ in diameter, but a few are just large enough to be visible to the naked eye.

At the opposite limit of the size scale are the viruses. Only a few of the very largest of them can be seen under the ordinary light microscope. One is the smallpox virus, which measures about 0.25μ. The others are so small that another new unit is required to record their miniscule dimensions: the $m\mu$ (a millionth of a millimetre). Foot-and-mouth disease virus, for example, is about $25m\mu$ in diameter. The electron microscope is essential in studying such creatures.

A virus is little more than a package of DNA (or in some cases RNA) which can be transferred from one living cell (whether of man, beast, plant or bacterium) to another. Viruses represent the ultimate perfection of the parasitic mode of life. They have no independent living existence (they can even be crystallized like chemicals and kept on a shelf in a bottle), but must invade living cells to replicate themselves. This process, when repeated sequentially, leads to the destruction of cells and can constitute disease. But viruses also act as agents of heredity, entering cells and causing changes that are usually not harmful and can be beneficial (either for the cells involved or for man). Bacterial viruses can confer on bacteria the capacity to produce or resist the action of colicins, offensive proteins deployed between different strains of bacteria. And virus infection is also the beneficent cause of many of the beautiful and variegated colours of tulips.

The sexual life of the microbe

Even in the apparently simplest microbes, sex life is rather more varied and interesting than that of *Homo sapiens*. In one form seen in bacteria — the nearest approximation to the procedure adopted by man and woman — there is direct contact between two cells, and a portion of DNA is transferred through a sex pillus. Microbiologists were shocked when they first discovered this act, because they had assumed that cells without a 'proper' nucleus could not enact a process akin to authentic sex.

In fact, this is only one of several sexual mechanisms used by bacteria. Another is the transmission of *plasmids*, genetic elements separate from the nucleus, which give their possessor such worthwhile skills as the ability to resist attack by antibiotics. Another method exploits the promiscuousness of viruses, which can ferry genetic material between donor and recipient. A fourth involves the direct transfer of free DNA from one bacterium to another. An organism capable of being the recipient partner in an act of this sort is termed 'competent'.

In other families of microbes, sexual procedures are more diverse again. Sexual arrangements are even used as a criterion for classifying the various groups. Among the fungi, for example, there are sexual and strictly asexual modes of reproduction, life cycles, fruiting bodies and many different types of spores. Most textbooks also carry a paragraph or two on what are termed *fungi imperfecti* — those fungi in which sexual relations have never been observed. In the fullness of time, however, it becomes embarrassingly apparent that the imperfection is in the observers, not the fungi. Many of the fungi placed originally in this dismal category have later been detected *in flagrante delicto*. A number have turned out to be merely transient stages in the life-history of more sophisticated forms.

Some fungi and algae go so far as to have distinct males and females. Microbiologists still have not become accustomed to this circumstance, and so need guidance in observation. As one substantial textbook carefully explains: 'In the most general view, a *male* is defined as an organism that produces a gamete that moves in some way towards a female gamete, the latter remaining more or less passive.'

Curious, this reluctance to recognize the genuineness of sexual conduct among the microbes. Coupled with our tendency to see them as lower orders, it implies more than merely a disinclination to believe in the seriousness of something so small. Could the reason lie in our subconscious realization that, unlike you and me and the rest of the human race, the microbes are potentially immortal?

Impressive metabolic versatility

Whatever the explanation, any sense of inferiority which we already feel towards the microbes is amply reinforced by even a casual inspection of their versatility, both ecological and metabolic. Consider temperature, for example. The human body has a comfortable temperature range of less than two degrees centigrade. And if our normal temperature (around 37 degrees centigrade; 98.4 degrees fahrenheit) rises above about 41 degrees centigrade (105.8 degrees fahrenheit) or falls below 32 degrees centigrade (89.6 degrees fahrenheit) we require urgent treatment. Bacteria do rather better. They have been found in boiling springs in New Zealand and Iceland, at temperatures of 92 – 100 degrees centigrade. They have also been recovered from the Arctic wastes at –12 degrees centigrade. Moreover, these are not simply temperatures bacteria can tolerate. They are temperatures at which they will grow normally. Probably no single microbe can thrive at both of these two extremes. But the temperature range for growth of any one organism (about 30 – 40 degrees centigrade) is still very much greater than for man or any other animal.

In many cases, the exotic extremes represent the normal environments for the microbes recovered: the organisms have not simply been transferred there by accident. Algae are often seen growing in such profusion on the surfaces

of snowfields and glaciers that, though individual cells
are not visible, their presence in astronomical numbers
imparts a red or green colouration to the surface. Algae
also live on the underside of ice floating in the Antarctic.
Despite the cold and the small amount of light that pene-
trates beneath the ice, they grow prolifically. At the other
extreme, hot springs in such areas as Yellowstone National
Park, Wyoming, are colonized by thermophilic (heat-
loving) algae and bacteria. Those with preferences for
different temperatures are to be found at different points
in the stream of water as it flows away from the spring.
From a human standpoint, the most surprising desire
shown by certain microbes, particularly bacteria, is to
be denied oxygen. Such *anaerobes* grow luxuriantly in
the absence of air. Many are even inhibited by its presence.

Other exotic locations whose conditions are not only
tolerated but actually relished by certain bacteria and
algae include Great Salt Lake, Utah, and the Dead Sea,
which contain a high concentration (about twenty-nine
per cent) of salt; and bogs, springs and lakes composed
of strong acid. Circumstances under which microbes will
merely survive are even more extreme. In 1968 scientists
working in the Antarctic came upon one of the huts used
by Sir Ernest Shackleton in 1913. Inside they found frozen
faeces, which they brought back to base, and bacteriologists
were later able to demonstrate that the samples still contained viable cells of *E. coli*. Research workers routinely
keep cultures at temperatures down to that of liquid
nitrogen (–195 degrees centigrade) — and this is actually
better tolerated than ordinary deep-freeze temperature
(– 20 degrees centigrade).

Microbes can withstand many other insults. Passing a
110 volt current (a.c. or d.c.) through a suspension of micro-
organisms appears to have no deleterious effect. Freeze-
drying is so inoffensive that it too is used to preserve
cultures of bacteria and other microbes for research. And
when several apparently unfavourable conditions have
been combined — as when microbes have been exposed to

simulated lunar or Martian conditions, for example, to assess the possible existence of life there—they have usually emerged totally unscathed.

The knowledge that micro-organisms can tolerate such bizarre conditions strengthens the belief that they may have played a fundamental role in producing the world's petroleum deposits. Pressure of up to 100,000 pounds per square inch are thought to have obtained in the depths of the oceans where the oil was first synthesized. Such pressures are well below those that robust microbial life can tolerate. Other likely constraints — salt concentrations of five to ten per cent, a temperature of up to 80 degrees centigrade and the absence of oxygen — would also be compatible with a microbiological process. Such a formidable set of conditions would, of course, at the same time rule out the possible involvement of other forms of life.

The tremendous environmental adaptability of microbes reflects their internal structure and chemistry. The enzymes of algae that live in boiling water, for example, must be very different from those in most other creatures. Enzymes are usually quickly inactivated when heated, so those found in these and other exotic microbes have to be unusually hardy. Similarly, there are special adaptations in the cell wall and membranes of bacteria that live in the presence of otherwise lethal concentrations of salt.

Metabolic tricks and specialities
An important factor behind the ecological versatility of microbes is the unrivalled variety of their biochemistry. All living organisms, micro-organisms included, share certain basic metabolic patterns — which is why microbes have proved such excellent material for scientific investigations into the 'secrets of life' (Chapter 9). But bacteria in particular are renowned for their metabolic tricks and specialties.

We and all other mammals gain our energy by oxidizing a strictly limited number of 'organic' substances — so

named because they are produced by other living creatures. Thus carbohydrates (sugars and starch) in the diet
have to be converted by the body to glucose, and that single
substance is then processed, through a stepwise series of
chemical reactions, to yield energy. Another pathway
leads from fat to the same sequence of energy-releasing
reactions. But the options are narrow. Many materials,
such as cellulose, which are in rich supply and offer an
excellent potential source of energy, are of no direct value
for man, because our restricted range of enzymes cannot
release the glucose they contain. Those species that *are*
capable of using cellulose as an energy source play host
to microbes that help them do so.

Bacteria are more talented than *Homo sapiens*. There
are bacterial species that derive their energy by oxidizing
such diverse 'inorganic' starting materials as ammonia,
sulphur, iron and nitrites. And while we need organic
foods to build our own tissues, as well as to acquire energy,
some bacteria can synthesize all their cellular requirements — a complex mixture of protein, carbohydrates, fat
and nucleic acids — from simple inorganic nutrients:
ammonia, carbon dioxide and water. Contrasted with
such ascetic needs, our own dependence on an entire
catalogue of nutrients, including amino acids (twenty
different sorts), vitamins, carbohydrates and fats, is somewhat unsettling.

The ultimate source of energy for most life on earth is,
of course, the sun. Some micro-organisms, including
algae and photosynthetic bacteria and protozoa, can use
solar radiation directly. They are *autotrophs*, deriving
their energy from the sun's rays and using carbon dioxide
gas from the atmosphere to build up their own organic
tissues. Other autotrophs exist by tapping some of the
energy released during chemical transformations of inorganic compounds. They are a mixed brigade, including
the bacterium *Beggiatoa* (which does a fine job by oxidizing
hydrogen sulphide — rotten eggs gas — into sulphur),
Nitrosomonas (which lives by converting ammonium salts

into nitrite in the soil and *Nitrobacter* (another soil inhabitant, which turns nitrite to nitrate).

Those microbes that, like man, have to use organic foods, usually command an infinitely wider set of options than we possess. Often they have the enzymatic capacity to utilize many different sugars and related substances. But they do not produce all such enzymes at any one time — only those they need to tackle the nutritional task in hand. Similarly, bacteria and their microbial colleagues fabricate a richly varied selection of end-products from their metabolism. It is this skill that we exploit in chemical industry and elsewhere, when we harness the chemical manipulations of the microbial craftsmen.

A particular strength of the 'lower' organisms is their absence of anything resembling a sense of smell. They will cheerfully ingest the most aesthetically displeasing materials as long as the basic elements are of the right type to suit their enzymatic abilities. There are species of bacteria and fungi that, given time, will consume such varied fare as railway sleepers, gas pipes, crab shells, animal hoofs and horns, faeces, old copies of *The Times*, grease and oil, leather, sawdust and rubber tyres. Few if any micro-organisms can tackle every one of these materials (though some come very close to doing so), but all of the commodities listed, together with thousands of others that we would find distasteful or poisonous, such as phenol, carbon monoxide, hydrogen and paint, can serve as food or part food either for a single species of microbe or for two or more species acting in concert.

How microbes evolved

Like some of us, a few microbes are extremely fussy about their diet. They need a large number of complex organic compounds, as we do, if they are to survive. But there is cold comfort here for those who might see reassuring kinship in this similarity, after so much evidence of the superiority of microbes. For the species that are most fastidious in their nutritional needs (and thus, it must be said, most vulnerable in times of food shortage) are almost certainly the most primitive representatives of their type.

This is another lesson that microbiologists have learned with uneasy reluctance in recent years. Formerly, they believed that the very earliest living forms on earth were self-sufficient autotrophs. Like present-day autotrophs, these were supposed to exist without organic foods. The idea was that, while some such organisms had persisted over the aeons of evolutionary time, others had gradually lost certain abilities to manufacture essential substances for themselves. Every time a metabolic capability was lost, that organism developed a corresponding dietary need for a substance it had formerly been able to manufacture. (This would parallel what is thought to have happened to vitamin C metabolism during human evolution. Primitive man, it appears, was able to synthesize his own vitamin C and thus did not require it in his diet. But this capacity disappeared at some stage, to be replaced by a corresponding nutritional requirement — and the appearance of scurvy when the vitamin was absent.)

Present-day views have turned this account on its head. According to the best evidence and soundest speculation, life seems to have originated on earth following an initial scenario of chemical synthesis. Starting with totally inorganic materials, relatively complex compounds — which we would now term 'organic' — were generated and they accumulated in the primeval soup. There was time — an abundance of time — for this painstakingly slow process, based on random, accidental reactions between simpler components of the ooze, to occur. And there were neither living creatures to consume the accumulating materials, nor oxygen in the atmosphere to oxidize them. Eventually, the first living, self-reproducing organisms came into being. They were heavily dependent on the rich stockpile of compounds in the environment. Only much later did the earliest creatures begin to build up their own metabolic capabilities — a process probably accelerated by the gradual depletion of the plethora of ready-made foods.

If this view is correct (and it seems to be), we must add nutritional inadequacy to mortality, metabolic dullness and sexual ordinariness as cardinal features that distinguish man from what he was once pleased to call 'lower organisms'.

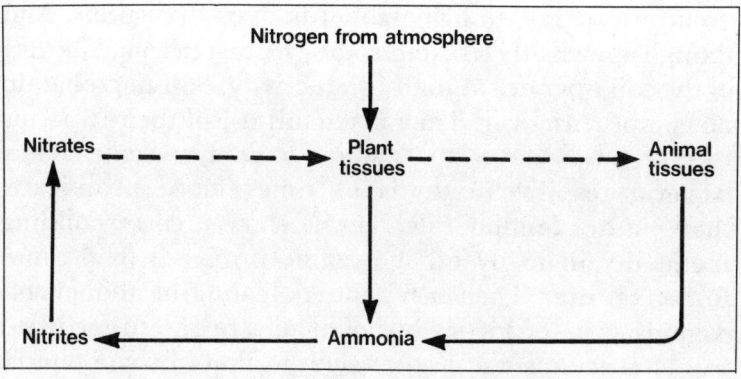

The nitrogen cycle. (See Chapter 3)
Solid lines represent conversions effected by microbes.

3 The Fount of Nourishment

The combined weight of all the microbial cells on earth is about twenty-five times that of the planet's animal life. By far the larger proportion of the micro-organisms making up this crude total are engaged in constructive, co-operative, healthy and wholesome activities – judged even from our own prejudiced viewpoint. Only a relatively trivial number are mobilized to produce and spread disease. Every hectare of well-cultivated land, for example, contains 300 – 3,000 kilograms of micro-organisms. They are involved in the vast and complex process of sustaining the organic economy. Co-operation, symbiosis and division of responsibilities between many groups, species and sub-species combine to recycle the earth's basic elements, replenish the soil and thus nourish the profusion of life as we know it.

Microbes use unique enzymes to effect these processes. They work quietly, giving no outward signs of the magnitude of the chemical transformations for which they are responsible. Chemical engineers would have to use fierce conditions – considerable temperatures, perhaps, or great pressures – to bring about the same reactions. And though apparently so lifeless, the global, microbial factory of the soil operates at high productivity. Soil microbes do not simply tick over. They make full use of their superior rate of metabolism. In relation to body weight, many bacteria respire hundreds of times more prolifically than we do. Similar calculations show a corresponding inequality in many other organic processes in the two forms of life. The metabolic potential of the micro-organisms in the top six inches of an acre of well-fertilized soil at every instant is equivalent to that of some tens of thousands of human beings.

A Russian, Serge Winogradsky, and a Dutchman,

Martinus Beijerinck, were the two great figures who first brought to light the cardinal significance of micro-organisms in the biologically crucial cycles of matter on earth — principally those of carbon, nitrogen and sulphur. They showed that the extraordinary metabolic diversity of microbes was essential in promoting the recurrent recycling of these elements between plants and animals — both of which have strictly limited metabolic talents.

It was Winogradsky who discovered the autotrophic bacteria, which play a pivotal role in this inexorable process. Winogradsky and Beijerinck shared the credit for the other major discovery, that of the fixation of nitrogen in the atmosphere into a form in which it can be used by other living creatures. They found that certain blue-green algae and bacteria — some living freely, others associated with plants — are alone responsible for bridging this major gap in the organic economy.

Beijerinck and Winogradsky were wise men. They saw that laboratory techniques in science were to be used with prudence, and should not be allowed to dominate one's thinking. In this they diverged from many of their con-temporaries working with disease-causing microbes. For practical purposes, it is useful and even essential to isolate what are called pure cultures. A swab from the back of a patient's throat, or a sample of soil, will contain a myriad different types of microbe, and researchers have devel-oped 'enrichment' methods to select the particular strain in which they are interested. Laboratory studies are then carried out with the pure culture.

This has led many medical microbiologists to neglect microbial ecology and diversity, and to imagine them-selves as battling against single, defined enemy strains. Winogradsky and Beijerinck recognized the value of methods for separating individual bacterial strains from mixed populations; indeed, they devised techniques that are still used today. But they were primarily con-scious of the need to study the natural, mixed ecological communities of microbes in the soil and elsewhere. Both practically and in research, this is an injunction we neglect at our peril.

The story of nitrogen

The cyclic transformation of nitrogen is of paramount importance to life on earth. Nitrogen is an essential element as a constituent of nucleic acids and amino acids, the building blocks of proteins (including enzymes). Yet, while it is the commonest gas in the atmosphere (comprising about eighty per cent of air), animals cannot use it in that form. Neither can any but a few species of plants — and they require the assistance of microbes in doing so.

The task of the nitrogen-fixing microbes is to secure nitrogen from the air and combine it with other elements to form organic compounds in living cells. When those cells later die, the nitrogen, now in a fixed form, is available to plants, either directly or after further microbial manipulations. We, in common with all other animals, have to receive all our nitrogen in a fixed form by eating plants or animals that have eaten plants.

One of the principal nitrogen-fixers is *Azotobacter*, a soil bacterium that thrives in well-aerated and neutral or slightly alkaline arable soils. It is almost an autotroph in its life style, though it does use some organic materials. These come from the decomposition of cellulose, starches and similar plant constituents by other soil microbes. Carbohydrates such as starch, added to the soil, stimulate *Azotobacter* to grow and thus increase the fixation of nitrogen.

Biologists once supposed that the contribution of *Azotobacter* to nitrogen fixation was relatively modest, but estimates have been successively upgraded in recent years. So too has that for the amount of nitrogen harnessed by blue-green algae in the soil. This is particularly substantial in the tropics, where as much as thirty-five kilograms of nitrogen is fixed per acre each year. The size of the contribution by the blue-green algae can be judged from the fact that many of the world's paddy fields, where they are particularly abundant, receive no artificial fertilizer whatever. Yet over half the world's population lives on rice as a staple diet. The nitrogen seized from the air by these various fixers is made available, as organic compounds, in the form of secretions and excretory products — and in the organisms themselves if and when they die.

On a world scale, even more important nitrogen-fixers are the species of *Rhizobium* that live inside nodules on the roots of leguminous plants such as clover, alfalfa and soya beans. Man has long recognized that by rotating crops, one of which is a legume, soil fertility can be maintained. Conversely, fertility declines if the same plant, whether grass or barley or wheat, is sown year after year. The explanation lies in the fixing of nitrogen by symbiotic bacteria in the nodules. The bacteria capture more than enough for their own purposes — or even for the needs of the host plant. They are so efficient that even when legumes are planted on soil deficient in nitrogen, the net amount of nitrogen in the soil increases. It was Beijerinck who proved that legume seeds lacking the bacteria — which would therefore not nodulate — produced the characteristic nodules once more if treated with cultures of *Rhizobium* that had been isolated from other leguminous plants.

Many of the nodule bacteria are specific to their natural partner. Those from peas, for example, will not provoke the formation of nodules in lupins, or vice versa. Though the bacteria can often be found in soil, therefore, there are occasions when a leguminous crop does not develop nodules normally, because its special bacteria are absent or deficient in the soil. Whereas the soil under a crop such as wheat may have less than ten *Rhizobium* cells per gram, the same soil will carry between 100,000 and 10,000,000 cells per gram after the establishment of a flourishing legume crop.

As with *Azotobacter*, the fixed nitrogen is built into amino acids and other nitrogenous compounds. Here they are found in the bacteria, the plant and the surrounding soil. The relationship is one of genuine symbiosis. Nutrients from the plant nourish the bacteria, which in turn secure more than enough nitrogen for their own use. Neither the bacteria nor the legumes can capture nitrogen by themselves. For man, the net value of this co-operation is clear: a well-nodulated crop of alfalfa introduces as much as two hundred kilograms of fixed nitrogen per acre per season.

Despite its generous contributions, nitrogen fixation by symbiotic bacteria is still not fully understood. When microbiologists first examined the nodules of leguminous plants, they diagnosed them as pathological aberrations, rather like the galls induced in some plants by infestation with insects. It was only in 1962 that microbiologists finally convinced themselves that the root nodules found on alder plants also carry bacteria that fix nitrogen. Even today, at a time of acute food and fertilizer shortage, our understanding is incomplete. There are, for example, numerous other nodules, of unknown function, on the roots and leaves of plants, which may well harbour bacteria active in fixing nitrogen.

Typical of our still developing knowledge of nitrogen fixation was a surprising discovery announced towards the end of 1973 by research workers at the University of Wisconsin. It concerns the white-ant, or termite, long regarded as one of the world's most unwanted nuisance species. Now, it seems, as well as playing an important role in breaking down dead trees and other forms of cellulose, worker and to a lesser extent soldier termites are in the business of nitrogen fixation. Both activities rest on another, symbiotic relationship with microbes. According to Dr John Breznak and his colleagues, the termites play host not only to bacteria capable of digesting cellulose but also to those that fix nitrogen.

The bacteria live in the termites' gut and their presence seems to explain the remarkable ability of the insects to survive with only cellulose as a food material. (Some termites can live on a diet of nothing more than filter paper.) While both termites and bacteria derive something from their relationship, the main beneficiary is the soil. It is replenished with nitrogenous substances and cellulose breakdown products, and its aeration and drainage are promoted by the energetic activities of the termites. Another intriguing discovery, announced in mid-1975, concerns the ability of certain Brazilian grass and even maize strains to fix nitrogen by virtue of the *Spirillum* bacteria they carry. First noted by Dr Johanna Döbereiner, this phenomenon occurs only at comparatively high temperatures and its significance in nature is not yet apparent.

If all nitrogen, once fixed, were to remain inextricably bound up in organic matter, it would be pointless to apply manure or any other dead animal or plant material to the land. In reality, once an animal or plant dies, it immediately starts to decompose, as does faecal and other waste. The ubiquitous microbes begin their work of meticulously dismantling the abundance of large and complex compounds in the dead tissues, converting them to simpler, smaller molecules that are then returned to the soil. Many different micro-organisms play their part in this operation, which, in the case of nitrogenous substances, yields ammonia as its chief end-product. Urea, a waste material in human and animal urine, is also attacked by various bacteria, to yield ammonia. (Hence the smell of babies' nappies that have been left too long without washing.)

So far, we have seen how microbial action captures nitrogen from the air, and then later releases that nitrogen, in the form of ammonia, from the complex organic compounds in which it has been temporarily enshrined. Next, we must follow the ammonia, whose fate is also determined by micro-organisms. Here there are two distinct stages, and two highly specialized groups of bacteria do the work.

First, ammonia is oxidized to nitrite. Then the nitrite is oxidized to nitrate — the principal nitrogenous material available in the soil for the growth of plants. Several other bacteria and fungi also assist incidentally with this process of *nitrification*, but the chief bacteria governing both the first stage (*Nitrosomonas* and *Nitrocystis*) and the second one (*Nitrobacter*) are autotrophs. They conduct these chemical reactions so as to obtain energy to grow.

Thus the nitrogen cycle is complete, as plants use the soil nitrates to build up their own cellular structures. The cycle has a few additional, minor features, such as some loss of soil ammonia to the atmosphere. And of course there is the input of nitrogenous fertilizers synthesized artificially by the chemical industry. But even this contribution is puny compared with the concerted work of the trillions of microbes that sustain the natural nitrogen cycle throughout the world. Note, by the way, that we, humankind, are totally parasitic on the cycle. We rely entirely on organic nitrogen compounds whose synthesis was initiated by soil bacteria; and when we defaecate, urinate or die, our effete materials have to be reprocessed and recycled by further armies of micro-organisms.

We shall be discussing microbes as scavengers, breaking down and clearing away all manner of organic debris, in Chapter 7. Suffice it here to say that the folly of neglecting the natural, microbial activities of the soil is nowhere more apparent than in our crude misuse of fertilizers in recent years. To the blinkered chemist, replenishing the earth with large dosages of chemical fertilizers can more than replace the natural contributions of the soil microbes. To the agricultural engineer, intent on boosting productivity, the use of heavy machinery makes technocratic sense. In practice, the combination of both techniques has conspired to damage soil structure almost irreversibly in many parts of the world. Just as important as the chemical composition of the soil is its physical organisation and its flourishing organic life. Humus, a rich mixture of plant and animal remains being slowly degraded by microbial action, is crucially important. This soft, brown, spongy material is vital to the structure of the soil, making it friable. It holds water, like a sponge, and also acts as a reserve store for slowly released food for both plants and microbes. If this dynamic zone, or that around plant roots, is destroyed and if the soil becomes compacted by heavy machinery, then however rich it may seem when assessed by chemical analysis, the soil will become barren. In December 1974, research was reported showing that a delicate microbial balance exists in the soil, in which anaerobic bacteria producing ethylene gas play a crucial role. If disturbed, soil structure and nitrogen availability are impaired, and root disease increased.

Inexhaustible air?

As well as nitrogen, the two other elements vital to the existence of life as we know it are carbon and oxygen. Central to the cycling of these elements are the twin phenomena of respiration and photosynthesis. Respiration is the process by which animals and plants breathe in oxygen and use it to liberate energy from organic foods, producing carbon dioxide. In photosynthesis, plants consume carbon dioxide from the air and build its carbon into sugars, starch and other carbohydrates.

Again, the role of the microbes is one of massive support for the rest of the living world. Man, like other animals, depends absolutely for the replenishment of the atmosphere's oxygen on photosynthesizing organisms (which, although they may consume some oxygen, on balance give it out). On land, plants are the main contributors. But world-wide, despite the vastness of the forests and jungles, their share is the minor one. Marine creatures, principally bacteria and algae, yield the greater part of the world's oxygen. Well over 100,000 billion kilograms of glucose alone, and even larger quantities of oxygen, are generated in this way annually. The seaweeds (which are also algae) play some part, but they are confined to a relatively narrow coastal strip. It is the unicellular, microscopic algae, capable of developing wherever the environment is favourable, that contribute the major fraction. The importance of that share is emphasized by the fact that all of the oxygen in the atmosphere, were it not replenished, would be totally exhausted in under twenty years. (If that were to happen, of course, many microbes would survive perfectly happily without oxygen. Man would be doomed.)

Though the inexorable cycles of nitrogen, carbon and oxygen represent the most fundamental microbiological underpinning of organic creation, several other elements go through similar cycles. Here, too, microbes provide essential maintenance. Sulphur, for example, is a key constituent of several amino acids and other vital molecules, and is taken up by plants in the form of sulphates. These salts are produced by bacteria acting mostly on hydrogen sulphide, which is released during the decomposition of proteins by other microbes in the anaerobic process of putrefaction.

Further cycles include the phosphorous cycle, and the iron and manganese cycles. Particularly intriguing is the type of recycling that goes on in an anaerobic environment. In these days of newly recognized global scarcity, the concept of recycling has become popular even with the politicians, who could learn much by observing how the same principle is applied in the microbial world. In a typical microbial community adjusted to living without oxygen, photosynthetic sulphur bacteria use light and carbon dioxide to produce their cell material, simultaneously converting hydrogen sulphide to sulphate. When they die, their cell constituents are broken down by another group of microbes, releasing carbon dioxide, hydrogen and other products. These in turn are harnessed by other microbes that convert the sulphate back to hydrogen sulphide. And so on.

As the system is completely self-enclosed, it can be established in a sealed bottle that has been inoculated with the appropriate mixture of organisms from the mud at the bottom of a pond. Provided that the bottle is well illuminated, its microbial occupants will continue in their state of self-sufficiency and mutual support for several years.

Microbes and mutual aid

Two other models of symbiotic life styles should be considered. The first is the *mycorrhiza*, which means literally 'root fungus' and refers to the symbiotic relationship that exists between the roots of many plants and fungi. There are two types. In the first, which is characteristic of forest trees, especially oaks, beeches and conifers, the filaments of the fungus form an extensive sheath around the outside of the root. The second group, found typically in orchids, have the fungus penetrating deep inside the tissues of the root itself.

The fungi in tree mycorrhizae derive their nutrients from root secretions, and are never found in isolation in nature. Conversely, the fungus contributes to the tree's nutrition, and plant scientists have recently discovered that they help principally by promoting the tree's absorption of phosphorous from the soil. Pines absorb two or three times more phosphorous when mycorrhizae are present than when they are not, and many forest trees become stunted and die when they are deprived of their symbiotic fungi. Trees that are mycorrhizal, planted in poor soils, will thrive, whereas those lacking a fungal partner will not. The role of mycorrhizae in orchids is less clearly understood, though again it is thought to be mutually beneficial.

Another type of close relationship, not between a microbe and a 'higher' plant, but between two distinct sorts of microbes, concerns the lichens. It might be argued that a lichen — a regular association between an alga and fungus — is not of itself a micro-organism, because lichens can be seen everywhere with the naked eye, growing on walls, tree trunks, rocks and roofs. But by the same token we would have to agree that bacterial colonies, containing many millions of cells, growing on a microbiologist's culture dish were not microbes. The point is that, though in mass aggregates both lichens and bacteria are visible, their individual cells are not.

Lichen associations are particularly cosy. The algae photosynthesize (some fix nitrogen too) and produce for their own use organic compounds, some of which the fungi consume as nutrients. The algae in turn receive protection: a firm anchorage, shielded from the rain, wind and excessive light. The remarkable hardiness of lichens bears out the success of this arrangement. In many climes, such as deserts and polar regions, lichens are the dominant forms of life, and during winter are major sources of food for reindeer and caribou. They are important in helping to form the soil in some parts of the world by extracting minerals from rocks and forming humus.

In a bizarre way, the success of the lichen's way of life can be illustrated by the following clumsy sentiments from a contemporary textbook of microbiology: 'If a lichen is removed from its natural habitat and placed under favourable conditions (for example, with organic substrates or high humidity), the symbiosis breaks down; the fungus may proliferate and destroy the alga, or the alga may overgrow the fungus. . . . It is possible to cultivate separately in the laboratory algae and fungi, but it is difficult to resynthesise the lichen from the isolated components.' The author clearly has an odd view of what, from the standpoint of a lichen, are favourable conditions.

Whatever the inclinations of scientific experts to pull lichens apart, the creatures themselves are superbly hardy in nature. They are extremely resistant to heat, cold and desiccation, and often live in positions exposed to direct sunlight, where they are subject to intense variation in physical conditions. A lichen can tolerate total desiccation for several months, yet remain viable. When suddenly dowsed in water, as happens in a thunderstorm, it may increase its water content from two to three hundred per cent of its dry weight in just half a minute. Field measurements show that the temperature of lichens can reach up to 70 degrees centigrade. Conversely, they are remarkably unaffected by chilling, and have been kept at temperatures as low as −268 degrees centigrade, revived, and have resumed normal respiration afterwards. No other living creatures have such an impressive record of hardiness.

How to digest grass
The bulk of the carbohydrate in terrestrial plants occurs as cellulose — which is a form man and most other animals cannot use. The only known creatures with the digestive equipment to attack cellulose directly are microbes, such as *Cytophaga*, which produce enzymes called cellulases capable of doing so. The ability of the ruminant animals (including cattle, sheep, goats, camels and giraffes) to munch grass and leafy vegetation and derive nourishment from it rests on their harbouring, in special parts of their alimentary tract, massive cohorts of cellulose-digesting bacteria and protozoa.

Ruminants have four stomachs. The first two comprise the rumen and are teeming with organisms of this sort. In the cow the rumen has a volume of about one hundred litres, and it contains some ten billion micro-organisms per millilitre. Grass and other plant materials, amply mixed with saliva, pass into the rumen where they are efficiently attacked by the mixed microbial population and turned eventually into fatty acids, carbon dioxide and methane. The cow has to vent the gases by regular vigorous belching: some sixty to eighty litres are produced every day. The fatty acids are absorbed through the rumen wall into the bloodstream, and are used throughout the body as sources of energy.

As well as cellulose digestion, there is another ingenious aspect of ruminant metabolism. While man needs the so-called essential amino acids to make protein (a liability we share with the rat), ruminants can construct them from just ammonia or urea — end-products of excretion in many creatures. The reason is that the microbes in the rumen build up their own proteins from these simple substances. Every so often the rumen contents — partly digested food, together with some of the micro-organisms — pass from the rumen, which contains no digestive juices of its own, to the lower stomach, which secretes enzymes that digest protein. The material digested there includes not only protein remaining in the food, but also the microbes that came with it. Their nitrogenous compounds (and vitamins) are thus released and absorbed by the animal. Towards the end of 1974, research workers also reported that bacteria in the rumen of sheep fix nitrogen, to an extent that can be quite significant for sheep grazing on pastures low in protein.

The advantages of this situation to the animal are obvious. But what of the cohorts of bacteria and protozoa? They are provided with a virtually perfect environment. It resembles the 'continuous culture' apparatus, often used in laboratory research, in which the microbes receive a continuous flow of fresh nutrients, while all waste products are immediately swept away. Under these conditions, the organisms multiply at their maximum possible rate.

They are neither short of food nor poisoned by their own excreta — as happens in 'batch culture', where microbes grow in a restricted amount of medium containing limited nourishment.

Such is life inside the warm, anaerobic, churning environment of the rumen. The residents are surrounded by an abundance, continually renewed, of their favourite food, and they are kept at a constant, ideal temperature (39 degrees centigrade). Other physical conditions are optimal for their purposes. Most intriguing, the ruminant's salivary glands have been modified during evolution to provide a smooth balm for the ensconsed organisms. The saliva does not contain digestive enzymes. It is a dilute salt solution which provides a perfect, slightly alkaline medium within which the microbes can initiate their attack on the cellulose.

The complexity of ruminant digestion can be gauged from the fact that, when newly chewed food enters the rumen and mixes with the resident microbial population, it has to remain there for about nine hours for the necessary chemical processes to be completed. But the net result is one of impressive efficiency. Clearly, a ruminant is nutritionally superior to a non-ruminant when both are living on foods deficient in protein, such as grass.

A comprehensive account of ruminant digestion still has to be written. Only comparatively recently have research workers (principally Dr R. E. Hungate and his colleagues at the University of California) been able to develop techniques allowing the highly anaerobic conditions of the rumen to be stimulated in the research laboratory. What is simple for the cow is extremely difficult for humans to produce, even using sophisticated modern technology.

In nature, it seems, a comparatively small group of rumen bacteria create the strictly anaerobic state, in which oxygen is totally excluded, that allows the cellulose-splitting microbes to proliferate and perform. Formerly impossible to cultivate artificially, some of these have now been grown and identified. They have been discovered elsewhere, too. One site is the human gut, where they probably have an important role in suppressing the growth of pathogenic organisms.

As well as ruminants, some other animals have varying abilities to digest cellulose. Herbivores, including the rabbit, horse, guinea-pig, pig, rat and porcupine, can digest grass and leafy material through the good offices of microbes similar to those in ruminants. But here the organisms live in the caecum, a roomy region of the large intestine. Consequently, the whole process is less efficient. The caecum being below the stomach proper, microbial cells and part-digested food are lost in the faeces: they do not enter the stomach, as in ruminants. To some degree rabbits and guinea-pigs make good this deficiency by indulging in coprophagy — eating their faeces. Animals prevented from doing so frequently develop signs of vitamin deficiency and other indications of malnutrition.

Part of the motive behind recent research interest in ruminant digestion has been the possibility, in a world faced with a food crisis, of imitating ruminants and herbivores in their ability to convert herbage to high-quality protein. There have even been hopes of setting up vast tanks, like the ruminant, which could be fuelled with grass and inoculated with ruminant organisms, and from which nourishing substances could be removed at intervals. There now seems little prospect of being able to copy nature in this way — one of the chief problems is the enormous difficulty of maintaining, in a culture vessel, the totally anaerobic conditions that are achieved so easily inside the sheep or cow. As Dr P. N. Hobson, one of the principal research workers in this field, has commented: 'Considering all aspects together, it is probably best to leave rumen bacteria in the ruminant as a source of protein for humans.'

One of the most remarkable examples of microbes assisting 'higher' animals with their digestion is that of the honey guides, birds of the genus *Indicator*, which live in India and Africa. They are intriguing creatures, deriving their name from their habit of guiding both honey badgers and human beings to the nests of wild bees. Arriving on site, they wait for the 'higher' animal to break open the hive. Then, when he, she or it has left the scene, the honey guide settles down to feed on the remains of the honeycomb.

Beeswax presents a tough challenge to the digestive enzymes, and a few years ago scientists investigated how honey guides tackle this unusual diet. What they found was that the birds have no digestive enzymes to accomplish the task. Instead, they play host to two microbes, a yeast and a bacterium, which carry out the digestion for them. The microbes live in the intestinal tract of the honey guides, and one of them (the bacterium) benefits by receiving a substance it requires for growth and which is produced in the intestine.

Man's reluctant dependence

We have always been more willing to recognize the use that other animals make of microbes than to consider our own dependence on such inferior forms of life. The keeping of fungi by ambrosia beetles makes a whimsical story — but one that merely illustrates the occasional, accidental discovery by the lower orders of the merits of banding together in the struggle for existence. Another jolly example is that of those squids and fishes that harbour luminescent bacteria because, unlike fireflies and worms, they cannot manufacture luminescent chemicals in their own tissues. But the thought that we ourselves owe anything whatever to the microbes in our intestines, for instance, was stoutly resisted until very recently.

Look up any textbook of bacteriology published before 1970 (and some more recent ones) and you will find the bacteria that live in the human gut magnanimously listed as 'harmless commensals'. And the definition of commensalism is 'a relationship in which one species benefits from the association, the other being unaffected'. Like rumen microbes, the human gut inhabitants clearly gained from their occupancy of a warm and constantly favourable environment in which food was always available and wastes were regularly removed. But man could not imagine he gained any benefit from this arrangement, and could look only with benevolent equanimity on the ten billion or so bacteria that he disgorged in every gram of faeces.

So it was that, even in the 1940s and 1950s, when the most spectacular of the weapons in the modern armamentarium of antibiotics were being introduced and deployed in mass anti-microbial warfare, most doctors and scientists ignored the 'normal population' of the human intestine. As a result, many of the people given antibiotics, to treat a chest or skin infection, perhaps, developed gastric and digestive troubles, ranging from the trivial to the serious. The cause, when it was eventually investigated, was that the drugs, as well as killing their intended victims, were also slaughtering wholesale the microbes in the intestine. In some cases the gut was being virtually sterilized, with all vestiges of its normal population eradicated (and once this happens, it takes a long time to re-establish such a population). The lesson was clear. The microbial flora of the intestine was indeed important in human nutrition.

We are still working out the complex interplay between gut organisms that provides this support. But the recent change in microbiologists' attitudes is undeniable. Compared with the breezy certainty of even the mid-1960s, today's textbooks are notably modest on the subject. 'At present we understand only dimly why certain kinds of organisms are adapted to the intestinal environment and others are not', says one recent tome, authoritative on most other aspects. 'Even though we know much about the molecular biology of *E. coli*, we have no detailed idea of why its natural habitat is the intestinal tract.'

One type of research that is contributing to radical re-thinking of the role of the intestinal flora is *gnotobiology* — the study of animals raised in totally germ-free environments and those in which all resident micro-organisms are known and identified. Establishing germ-free colonies of animals is assisted by the fact that most mammals are sterile at birth, having been protected by the placenta until that time. If the foetus is removed aseptically just before the expected birth (during which the animal would receive its first inoculation with microbes from the mother's vagina and elsewhere), it can be transferred to a germ-free isolation chamber and maintained there in its primitive, microbe-free state. Using this technique, colonies of not only rats, mice and guinea-pigs, but also monkeys, lambs and pigs have been established free of microbes.

Although such colonies were created originally not to investigate nutrition but to study such problems as the development of the antibody-producing system, it soon became apparent that animals raised in the absence of microbes had different nutritional needs from those of conventional animals. One distinction is that they require vitamin K in the diet, unlike normal, germ-ridden animals. Vitamin K is used by the liver to synthesize pro-thrombin, which gives rise to the enzyme thrombin that directs the clotting process in blood. *E. coli* produces the vitamin in excess, and when this bacterium is introduced into the gut of germ-free animals and becomes established there, the symptoms of vitamin K deficiency — and the need for the vitamin in the diet — disappear. *E. coli* is clearly more than a passive commensal.

The vitamin K story applies to man as much as to many other animals. Indeed, babies are sometimes born that have not received an adequate amount of vitamin K from their mothers (perhaps because indiscriminate antibiotic treatment has curbed her gut flora). This makes them prone to haemorrhagic disease of the newborn. Daily dosages of the vitamin will help correct the condition. In an untreated case, haemorrhage may cause death within two or three days. If the infant survives that period, however, it will probably recover, because bacteria, newly established in the intestine, begin to over-produce vitamin K within a few days of birth.

The role of vitamin K synthesis by gut microbes is the most striking of the recently discovered contributions of those organisms to nutrition. Another is production of vitamin B12. Without microbial help neither animals nor plants can make this vitamin, absence of which leads to pernicious anaemia in man. Further contributions have been hypothesized or are suspected. We know, for example, that gut bacteria also synthesize other vitamins of the B group, and this could satisfy a significant proportion of the body's need. It seems too that the syndrome of lactose intolerance — a digestive imbalance caused by the body's intolerance of milk sugar — may be alleviated in some individuals by lactase, an enzyme produced by gut bacteria which breaks down the lactase.

The intestinal flora may be of the most direct nutritional significance when a person is unwell, or where the diet is less than adequate. Some recent research indicates that ammonia released when gut bacteria break down urea can be used in the synthesis of amino acids. Other studies imply that certain micro-organisms in the intestines may, like those in the soil, fix gaseous nitrogen in the form of amino acids whose absorption could help nourish the 'host'.

This is a comparatively new research area. We already know what drastic effects can follow the wholesale eradication of microbes from the gut. We have also discovered that even small changes in the balance of the intestinal flora can have a marked effect on a person's wellbeing. And a number of specific microbial contributions have come to light. But we still await even the beginnings of a comprehensive understanding of the complex and beneficent inter-relationships of the gut flora and thus of the degree to which we depend on our microbial fellows within.

A major technical obstacle in gaining this knowledge is the same as that which confounds the study of ruminant digestion: the infuriating difficulty of maintaining under laboratory conditions the anaerobic and other fastidious organisms upon which the process depends. Even when we can do this, a thorough understanding of the way the mixed population behaves will rest on the yet more intricate task of assessing the interactions between the various members of the heterogenous microbial community.

When we have such an understanding, it may be that by judicious manipulations we can even improve on the process. As Dr Marie Coates puts it: 'When the interactions between the consumer, his diet, and his gut micro-organisms are more clearly understood it should be possible to determine the type of flora most favourable to the host's wellbeing and to select the kind of diet likely to encourage its establishment.' That, of course, carries the comparatively new implication that we can gain most from the microbes not by eradicating them, by design or accident, but by supporting and sustaining them to the best of our ability.

4 Improving on Nature

The facts of the world's food shortage can be briefly and starkly summarized. The population of the planet is expected to reach some 7.2 billion by the end of the century, and even in the best years recently food supply has been increasing nothing like fast enough to feed that many mouths. Quite the reverse: famine is already commonplace across vast tracts of the earth; 460 million people suffer chronic malnutrition; and in 1972 the already desperate situation deteriorated savagely, with abysmal harvests in the USSR, China, India, Australia, the Sahel and South East Asia. For the first time in over twenty years, food output declined. Cereal production, which had been rising at twenty-five megatons a year, fell by thirty-three megatons. World food reserves slumped astronomically. Then in 1973 the energy crisis and the fertilizer shortage made matters even worse. Yet, at our present breeding rates, the earth spawns an extra seventy-five million people every year. That is about equivalent to the combined populations of Canada, Kenya, Burma and Peru.

No one should suggest that catastrophe can be averted by simple slogans or simplistic technical panaceas. (If such a lesson needed to be learned, it can be seen in the much-heralded Green Revolution, based on high-yield cereal strains. In the long term, this has yielded far less spectacular results than were originally envisaged.) What does make persuasive sense is the argument that the varied skills of bacteria, fungi, protozoa and the other microbes could make a massive contribution towards both our strategy and tactics in tackling the global problem with which we are confronted. In chapter 6 we shall be looking at novel uses of microbes themselves as food. Here we will consider ways in which the natural role of microbes in promoting and sustaining the growth of animals and plants — already, as we have seen, a massive and pivotal role — can be encouraged and thus harnessed further for human welfare.

Battle for wasted nitrogen

The air that overlies every acre of soil throughout the world contains about thirty-six million kilograms of nitrogen gas. Nonetheless, nitrogen is often the element that is in shortest supply in the soil in a form available for the growth of plants. The amount of fixed nitrogen in soil ranges from 5 to 250 kilograms per acre. Thus the fraction of the nitrogen in the biosphere that is immediately accessible to plants is only some 0.00001 to 0.001 per cent of the total.

There is substantial room for improvement. Already, one of the most important contributions that soil bacteriology has made to agricultural practice has been in revealing the relationship between legumes and their nodule bacteria — particularly in showing the importance of having the correct bacterial strain for a particular species of legume. This knowledge is now being applied routinely. Pure cultures of microbes are available commercially for inoculating legume seeds. Alternatively, the soil can be inoculated. Species of *Rhizobium* are introduced into virgin soils, or soils known not to be good supporters of legumes, before crops such as alfalfa or soya beans are sown. Once introduced, the bacteria are usually present indefinitely in the soil.

As a result of these techniques, clover yields have been doubled in some parts of the world, and alfalfa crops that would otherwise have been worthless have been saved. Financially, the returns are dramatic. Some twenty-five to one hundred kilograms of atmospheric nitrogen can be added to an acre of soil in a single year as a result of the turn-under of well nodulated plants. The cost of that amount of nitrogen as commercial fertilizer would be several thousand dollars.

Controversial results have characterized the other obvious area for boosting agricultural productivity by applying the insights of microbiology: inoculation of soils with *Azotobacter* and other free-living nitrogen fixers. It is an area ripe with possibilities. We need more insight into the ecology of soil populations before inoculation can be a routine success, but it seems certain that past failures have been a result of our crudity of approach.

The most interesting work of this sort to date comes from the Soviet Union and other Eastern European countries. In the USSR, 'bacterial fertilizers' are widely available and are said to produce a gain of about ten per cent in the yield of field crops, and to benefit some fifty to seventy per cent of the crops to which they have been applied. Among the plants reported to have responded favourably are cereals, potatoes, sugar beat, tobacco, cotton, tomatoes and several horticultural crops. The bacterial preparations — with names such as azobacterin — are often applied by spraying a suspension on to the seed. Otherwise they are included, in an actively growing state, in granulated peat that is added to the land when the seeds are sown.

Studies in Britain have not fully substantiated the beneficial effects of inoculation. Conversely, reports from Australia and India have confirmed that bacterial fertilizers sometimes increase the yield or promote the rate of plant development. The fact that experience generally has been so variable only confirms that more research is necessary before we understand the microbiological and chemical processes that follow inoculation — and before we can thus harness the technique more fully. The pressures arising from the food crisis and the fertilizer crisis have already provided the impetus for accelerated research in this field.

To add to the confused picture, some recent studies on *Azotobacter* inoculation suggests that the microbe works, when it works, not through nitrogen fixation, but by

producing growth-regulating substances called gibberellins (p. 62). Dr Margaret Brown and her colleagues at
Rothamsted Experimental Station, Harpenden, have
found that, though the final yield of tomatoes may not be
affected after inoculation with *Azotobacter*, early growth
and flowering are both advanced. Other researchers have
reported that inoculation hastens flowering of wheat and
the growth of stems in potatoes and tomatoes. This suggested to Dr Brown and her colleagues that formation of
plant-growth stimulants, rather than nitrogen fixation,
was the explanation. This is now virtually certain. The
fact that the mechanism behind (one of) the favourable
effects of inoculation has proved different from that
expected does not, of course, render the work of any less
potential practical value. If anything, it emphasizes the
need for further research into the many different and
positive effects already achieved.

Science that is interesting and *useful*
Nitrogen fixation is a prolifically growing research area.
It is a splendid example of science in the anxious world
of the 1970s, being both a fascinating topic in fundamental
research (how is it that microbes accomplish quietly and
without fuss a reaction that otherwise needs a molten
lithium or red-hot manganese catalyst?) and one likely
to bring tangible practical returns. Although, in the long
term, one hopes that population control will reduce or at
least stabilize our global demand for food, in the imme-
diate future any steps that can be taken to increase nitrogen
fixation are to the good.

True, there are chemically synthesized fertilizers. But
as well as being increasingly costly, and thus often beyond
the means of the countries that need them most, they cannot fully replace the rich, organic activity of natural fertilization. In addition to the crumbling of soil structure that
is the inevitable sequel to their heavy use, imbalanced
application of synthetic nitrogen fertilizers can easily
precipitate deficiencies of other elements. In many parts

of the world during the past decade, sulphur deficiency has become apparent in soils, to such an extent that the sulphur cycle, rather than the nitrogen cycle, has emerged as the critical factor limiting plant growth. Microbiological fertilization, encouraged with ecological wisdom, would never have such bizarre consequences.

Looking to the future, one of the genuinely bright prospects is the creation of new nitrogen-fixing plant species or hybrids. Two possibilities suggest themselves. One is that of establishing a strain of *Rhizobium*, or another microbial fixer, in a crop plant such as wheat, oats or barley — combining, as it were, in a single year two of the crops of a rotational system. Experiments have not yet been totally successful in persuading non-legumes to accept a bacterial partner. But very recently plant scientists have succeeded in encouraging new symbioses between *Azotobacter* and carrot and other cells. With fuller understanding of symbiosis in those plants where it does occur, we may be able to arrange productive marriages of this sort — once we know what are the acceptable conditions for both parties. It may be that this will necessitate applying the emerging science of 'genetic engineering' (p. 225) to alter the genetic makeup of the bacterium and thus make its presence tolerable, if not welcome, to the plant. Alternatively, genetic engineering may facilitate the direct transfer to the plant of those bacterial genes that control nitrogen fixation.

As we saw in the last chapter, nitrogen is not the only element whose continual circulation throughout the planet is totally dependent on microbes, though it is arguably the most important. Some of the other transformations and translocations effected by microbes have also been boosted in recent years by human volition. The pine plantations of Puerto Rico, for example, failed until 1955, when mycorrhizal fungi were introduced, and they have been successful every year since. Similarly, preparations containing bacteria that make phosphates and other minerals soluble have been developed for inoculating into seeds before sowing. Microbiologists now feel that only a very primitive start has been made with such techniques, and that there is great scope for further development.

Foolish seedling substance

By far the most dramatic plant stimulants yet known to be manufactured by microbes — in this case fungi — are the gibberellins. Their recognition came originally from a study of *bakanae*, the 'foolish seedling' disease of rice, so named because affected plants first grow very quickly and then die. Known for at least a century as an occasional cause of serious damage to rice crops, it was first fully investigated in the 1920s by a Japanese microbiologist, Dr E. Kurosawa, working in Formosa. He found that the fungus responsible for the disease, when grown artificially, produced a substance that stimulated the growth of rice seedlings just as dramatically as did the fungus itself. The effect was of obvious potential agricultural interest, and scientists at the University of Tokyo tried to isolate and identify the active chemical. By 1938 they were able to publish a paper in which they announced that they had crystallized it. Named gibberellin, their material later proved to be a mixture of several closely related substances.

Sociologists and philosophers of science make much of the internationality of the scientific research community. Sharing a common conceptual structure, and resting on mutual trust and open comparison and criticism of new ideas, science can indeed bring together in common cause people from very different parts of the world, much as religion does. But there is still the language problem. So it was that, though published in the open literature, and despite its enormous practical potential, the early highly promising work on gibberellins went almost totally unnoticed in the Western world. It was, of course, published in Japanese.

Over a decade passed before scientists in Britain and

the US began to study the new group of growth stimulants. They soon found that virtually all plants respond to treatment with gibberellins, provided the substance can penetrate into the growing parts. Although only minute amounts are required (0.001 to 1.0 parts per million), the effects become apparent very quickly — within a few hours in herbacious plants. Gibberellins increase both the enlargement and the division of cells.

The early, sensational claims for the value of gibberellins, which hit the headlines during the late 1950s and early 1960s, have been modified by later experience. Some of the plants whose growth is handsomely stimulated by gibberellins lose succulent taste or other qualities. Others do not respond particularly well. Nonetheless, gibberellins have become widely used in agriculture, horticulture and malting. One of them, gibberellic acid, is applied on a considerable commercial scale for treating lettuce, rhubarb, pears, oranges, lemons, grapes, cherries, artichokes, cucumbers, celery and winter spinach. With vegetables the chief rewards are in better yields and earlier marketing, while gibberellin treatment of fruit trees leads to improved quality of the product or increased fruit 'set'. A different type of gibberellin shortens the period of development of biennial plants, so that seed crops can be obtained from lettuce and sugarbeet in the first year instead of the second. Another application is in reducing the dormancy of potato tubers, by dipping them in a gibberellin solution. In malting, the inclusion of gibberellin in the steep water hastens barley germination and also heightens the quality of the malt.

Gibberellins are produced industrially by growing huge quantities of *Gibberella fujikuroi*, the same microbe that causes foolish seedling disease. Manufacturers use specially selected and improved strains of the fungus. But it is notable that here, as in many other microbial processes, we continue to depend on natural microbes for making a substance whose mass synthesis is far beyond the skills of the chemical engineer. The manufacturing technique, similar to that used to make penicillin, is based on cultivation of the fungus in submerged, aerated conditions in giant fermentation vessels. And nothing need be wasted in gibberellin manufacture: the waste filaments of fungus left over at the end can be incorporated into feedstuffs for cattle.

As often occurs when initial dramatic boasts for a new product are not fully borne out in practice, there was a tendency after the early enthusiasms to write off the gibberellins as far less important and valuable than had been supposed. Only today is a really hard-headed re-assessment of the great variety of their potential applications taking place. There is, for example, increasing interest in treating many crops with gibberellins as a means of facilitating mechanical harvesting. It seems that, by altering growth habit and structure, or by compressing the harvesting period, the losses that are normally inherent in machine harvesting can be greatly reduced. Another proposed use is in promoting the establishment and early bearing of new apple trees, and perhaps altering the shape of the fruit. Field crops, grass, hops and sugar cane are also showing improved yields as a result of gibberellin treatment. One of the lessons here is that it takes time to learn the precisely correct way of using growth promotants for maximum effect.

Microbes against pestilence

'Wherever man grows food, insects are there to profit by his labours, and wherever he stores his crops, he will find insects are waiting for their share.' Those were the opening words of a superb film made by the Shell Film Unit in 1955, at the height of an enthusiastic boom in the development and deployment of chemical pesticides. Everyone then believed that the mass use of pesticides would win back for man the monstrous proportion — about a third overall — of the world's crops that were being destroyed by insect attack.

Those hopes have not been fulfilled. Despite the dedicated and ingenious labours of entomologists and industrial chemists, and despite their success in some of the battles along the way, the insects have won the war. Over three hundred species of insects have already become resistant to the chemicals, particularly DDT and the organochlorines, upon which so much faith was placed. By the inexorable process of mutation and Darwinian selection, the ability to resist attack by a particular insecticide can quickly emerge as the common characteristic of not only one population of insects, but the entire species.

Man's response has usually been the not particularly intelligent one of wielding the chemicals in bigger quantities, or applying them more often (with commensurate increases in cost). Sometimes this works. Resistance is not an absolute, black-and-white property. An insect that can withstand a modest concentration of DDT may well succumb to a dose ten times as heavy. But such tactics only delay the inevitable. Resistance to steadily higher levels of pesticides has been the pattern observed repeatedly in campaigns to destroy insect pests by chemical overkill. A side effect of this escalation has been to increase the levels of pesticide residues in the biosphere, which introduces considerable and often unpredictable environmental dangers. The use of DDT has been drastically curtailed for just this reason.

The end result, then, is that man as chemist has become virtually impotent to deal with the insect pests that ravage his crops — and do so most cruelly in just those parts of the world where the crops are most desperately needed to feed the starving. In the cool, factual words of a special report on the subject issued by the World Health Organization in 1973: 'At present, resistance in some species has developed to such an extent that chemical pesticides no longer give economic and safe control.' The practical reality behind that statement — illustrated by countless different, dismal setbacks throughout the world — is that agriculturalists have had to abandon any hope in chemical warfare on pests.

There is another way. It is called biological control, and it means exploiting the natural, ecological relationships that already obtain between living organisms, rather than crudely discharging toxic chemicals into the biosphere. Several of the rod-shaped bacilli, for example, are insect pathogens, and in recent years agriculturalists have begun using them to attack the agents of pestilence. The bacilli concerned are unusual in producing protein crystals when they form spores. They cause fatal disease in the larvae (caterpillars) of a wide range of insects, including the cabbage worm, the gypsy moth, the tent caterpillar and the silkworm. Some years ago microbiologists succeeded in isolating the protein crystals, and found that all the symptoms of the natural disease could be reproduced by feeding larvae on leaves coated with the purified crystals.

Bacillus thuringiensis has been particularly closely studied. Its crystals are diamond-shaped, and as little as half a microgram is sufficient to cause paralysis in susceptible larvae. The crystals are insoluble in neutral or slightly acid conditions, but dissolve in dilute alkali. The gut contents of most insect larvae are alkaline, so that when a larva consumes some of the crystals (or the whole bacilli) they dissolve when they reach the intestine. The dissolved protein then attacks the cementing material holding the cells of the gut wall together. The wall thus becomes permeable, and the alkaline liquid in the gut passes into the bloodstream. The result is generalized paralysis, followed some time later by death. At least 130 different species of moths are susceptible to the crystals. Those that are unaffected do not have a particularly alkaline intestine, so the crystals pass through without dissolving.

An agent of this sort has many advantages over a chemical as an insecticide. The most important is that, apart from the insect species that they destroy, the crystals are harmless to other animals and plants. Within the last decade recognition of this fact, together with the alarming failure of chemical pest control, has led to the rise of a new industry: the large-scale production of *B. thuringiensis*, which is incorporated in dusting powders used to protect commercial crops from insect destruction. The bacilli are grown in large vessels, like those employed in gibberellin manufacture. When a sufficient mass has been obtained, they are induced to form spores, which are then dried and incorporated in dusting powder.

In the US manufacturers produce several hundred thousand kilograms of powders containing *B. thuringiensis* each year, and production is rising quickly. There is a strict system of registration, based on proof that the preparation is both a safe and effective control agent. Currently registered uses in the US cover more than twenty agricultural crops and insect species, including the cotton bollworm, the cabbage looper (which as well as cabbage also affects beans, broccoli, cauliflower, celery, cucumber, kale, lettuce, melons, potatoes, spinach and tobacco), the European corn-borer, gypsy moth, California oakworm, fruit-tree leaf roller and the tobacco budworm.

Most insecticides of this type do not persist from year to year. That, indeed, can be considered an advantage, contrasting with many chemical pesticides. They have, however, to be applied very efficiently to ensure that the entire crop is covered. As so often in the microbial world, there is a multiplicity of different varieties. Some *B. thuringiensis* strains are lethal for a wide range of insects; others act only on one or a few species. Current developments suggest that there will be an increasing tendency to produce and market a variety of different forms for use against individual pests.

The first major successful attempts to combat an insect nuisance by biological control were based on the 'milky diseases' of scarabaeid grubs in the US. The best known of these diseases (so named because of the characteristic milky appearance of the dying larvae) is that of the Japanese beetle. The bacillus responsible, *Bacillus popillae*, has been employed with striking success in control campaigns.

Japanese beetle was introduced into the US some years ago, and it became a pest on pastures, shrubberies and lawns throughout a large part of the New England states and Canada. The larvae feed on the roots of grasses and other plants during the late summer, and it is at this stage that they can be attacked by the offensive bacillus. After an intermediary pupal stage, the adult beetles emerge in about the second week of June. They live for some forty days, during which time each female lays up to sixty eggs.

In contrast to *B. thuringiensis*, with its protein crystals, the details of the way *B. popillae* attacks its victims are obscure. What we do know is that the spores germinate in the gut, giving rise to a massive multiplication of bacteria inside the insect. Another difference is that we have not yet learned how to cultivate the spores artificially in the mass. 'Manufacturers' still obtain material for use in making their spore powders by collecting infected larvae in the field — a method pioneered by the US Department of Agriculture as far back as 1939.

The spore powder is applied to the soil, and the disease then travels through the population, carried partly by infected insects, but also by wind and water and by birds. After treatment, there is usually a spectacular fall in the number of Japanese beetles in the area, the disease spreading throughout the entire population within three seasons. The milky-disease bacteria are thought to have great potential for use against other insect pests, and trials are now in progress to establish their further versatility.

After their successes with two bacilli, microbiologists have been turning their attentions to other microbial pathogens that might be exploited in the same way. Over ninety different varieties of bacteria have been isolated from insects, in addition to which many others are suspected of being pathogenic. There is clearly a potentially invaluable lode here, only a very tiny part of which has yet been fully investigated.

Another topic receiving increasing attention is the possible use of fungi to combat crop pests. Fungi are attractive as control agents because they act as 'contact insecticides'. They do not depend upon the insect consuming them, but invade their victim's tissues from the outside. The drawbacks are that this is a slower process than with the toxic bacilli, and that its success depends to a greater degree on environmental factors, particularly temperature and humidity. This may well explain the marked contrast, ranging from decisive success to dismal failure, in the reports of recent field trials using fungi for biological control.

What of resistance? With more intensive deployment of microbes in insect control, must the emergence of invulnerable strains — a problem that has bedevilled chemical control — loom on the horizon? In practice, there are no instances yet on record of such resistance arising — despite the dissemination on a very wide scale of such bacteria as *B. popillae* against Japanese beetle. That may be no more than a reflection of the comparatively short time during which microbial insecticides have been in use. We should remember, however, that one of the great advantages microbes have over chemicals is that they are themselves living, and thus subject to the same natural processes of mutation, adaptation and evolutionary change that allow resistance to arise and resistant strains to proliferate among the insects.

One opportunity open to the biological control industry that is not available to the chemical industry, therefore, is to modify and change microbial agents with almost infinite subtlety in the laboratory. This can already be done by conventional breeding methods which utilize the known sexual habits of bacteria. In due course, it may be achieved by applying the technology spawned by the emerging science of genetic engineering. Already, commercial manufacturers of *B. thuringiensis* have used this opportunity to their advantage. By exploiting mutations and selecting strains for increased virulence, they have boosted the toxicity of their preparations considerably over the past few years.

Lessons of myxomatosis

The importance of understanding thoroughly how a pathogenic microbe and its host interact when the two populations meet on the grand scale is underlined by the saga of myxomatosis, introduced deliberately into rabbits in Australia and Europe in the early 1950s. Modern knowledge of this ghastly virus disease—characterized by grotesque mucinous tumours on the skin—dates from just before the turn of the century. In 1895 the government of Uruguay invited Guiseppi Sanarelli, a microbiologist at the University of Siena, to set up an institute of hygiene at Montevideo. Sanarelli accepted and in the course of establishing the institute he introduced European domestic rabbits to Uruguay for the first time. They were needed for producing sera containing antibodies against various diseases. The following year the rabbits developed the extremely infectious and lethal disease, then unknown in Europe, which we now call myxomatosis. Nearly half a century elapsed before the common wild rabbit of Brazil was incriminated as the carrier of the virus, and the method of transmission, via mosquito bites, was firmly proved. Mosquitoes also convey the disease in Australia and parts of Europe, but rabbit fleas are the major vector in Britain.

Because the European wild rabbit is a major pest in Australia, agriculturalists tried repeatedly to establish the myxomatosis virus there. But success did not come until 1950, when infected rabbits were introduced into Murray Valley. The splendid weather that summer and in the following two years provided unusually favourable conditions for mosquitoes to breed and travel far, and myxomatosis spread like wildfire. Millions of rabbits — about four-fifths of those in south-eastern Australia — succumbed.

Myxomatosis was introduced into France in June 1952,

and by the end of the following year it had reached Belgium, Luxemburg, Germany, the Netherlands and Spain. The first outbreaks in Britain were in Kent, and the disease moved so rapidly that by the end of 1955 well over nine-tenths of the rabbits in the country were dead.

Yet Britain's rabbit population has begun to thrive again in recent years. The rabbit is again considered a serious pest and a significant factor in agricultural economics. In Australia, though a myxomatosis epidemic occurs most summers, the disease now has only minor significance in regulating the rabbit population. What is the explanation for this dramatic change? And what can we learn from it?

The first, and unexpected, alteration was in the virulence of the virus. The lethal microbe first introduced into Australia killed over 99 per cent of infected rabbits. Yet within twelve months, strains of virus had appeared with a mortality rate of only 90 per cent. In subsequent years even milder strains have prospered, in some cases having mortality rates as low as 20 per cent. Obvious with hindsight, the outcome is one that nobody predicted at the time. The explanation is simply that, from the point of view of a virus, there is little point in destroying your victim in spectacular fashion if this leaves you high and dry, with no prospect of being ferried to other vulnerable hosts. The chances of further transmission are much greater for a virus that is comparatively mild, and which therefore produces a lengthier disease and infectious period, than with one that kills its host rapidly.

Rabbits have changed too. Under the rigorous pressures of regular exposure to a virus that was lethal to over 90 per cent of the population, resistant strains have emerged. During a seven-year period in Australia resistance increased to such a degree that a virus that originally killed 90 per cent of wild rabbits would eventually destroy only some 30 per cent or so.

The lessons are clear. Firstly, the conditions for successful microbiological control are so complex that initial

failures need have little significance for the eventual outcome of a project. Secondly, genetic changes in both the
microbial and host populations can quickly alter the
ground rules — though the extent to which this happens
will depend on environmental factors and on intelligent
anticipation. Above all, the long-term success of biological
control rests on a thorough understanding of the spread
and persistence of disease in natural populations. At present, that understanding is far from complete.

Virus versus the sawfly
All of this is relevant to the most rapidly expanding field
of microbial control: the exploitation of the great variety
of virus diseases in insects. Two main groups of insect
viruses, out of three hundred or so known to exist, show
high promise as insecticides. They are the nuclear polyhedrosis and granulosis groups. One of the former was
the agent for the first unambiguously effective example of
pest control by a virus, during the 1930s. At that time the
European spruce sawfly had been introduced into Canada
and parts of the US. No longer subject to its natural predators, it spread rapidly and became a considerable nuisance, destroying thousands of acres of spruce trees.

The ravages of the sawfly were checked somewhat by
the chance arrival of a nuclear polyhedrosis virus from
Europe. But another strain of virus, consciously introduced in 1959, has been so effective that it has replaced
the chemicals that were formerly used to combat the insect.
The amount of virus required for effective control is stag-
geringly small. It has been not only calculated, but confirmed in practice, that the virus from only fifteen diseased
pine sawfly larvae, mixed with a gallon and a half of water,
is sufficient to spray an acre of young pine trees, thereby
establishing an infection that controls the pest.

The present phase of feverish interest in virus insecticides was precipitated not only by the alarming decline
in the efficacy of chemical fertilizers, but also by the discovery by US Department of Agriculture research workers

of synthetic diets suitable for rearing the insect larvae as a source of virus. Using this technique, large quantities of virus have been prepared. Already, field trials have indicated great success in using nuclear polyhedrosis viruses to attack cotton bollworm, the tobacco budworm, cabbage worm and armyworm. The performance of the virus against the bollworm, as assessed by cotton yields, is about the same as that achieved by chemical methods. This, of course, implies a substantial advantage. As well as having no ill effects on plants or animals, leaving no offensive or harmful chemical residues and being specific to the pest it is designed to attack, the virus does not kill other natural predators of the pest — another disadvantage of many chemical pesticides. Similar success has been reported from the use of viruses to control coconut rhinoceros beetle on islands in western Samoa.

In these times when we are infinitely more conscious— rightly so—of environmental damage and of the unforeseen side effects of new science and technology, virus insecticides are being scrutinized very much more thoroughly than were many chemical pesticides when they were first introduced. In fact, all the evidence indicates that they are conspicuously safe, to man and indeed to all species of life other than those for which they are intended. Perhaps the most graphic illustration of this comes from the making of coleslaw. When the cabbage looper caterpillar falls prey to a virus, its body dissolves and sheds on to the leaf vast quantities of virus. The virus is not killed during any of the stages in the making of coleslaw. This means that during the late autumn in the US, when mortality among the hoppers is at its height, an average bowl of coleslaw contains some four hundred billion live particles of cabbage looper nuclear polyhedrosis virus. It would surely be apparent by now if the virus were harmful to man.

Moreover, if cabbage looper virus were to be used widely as a virus insecticide, there would probably be fewer viruses in the coleslaw, not more. This is because the main result of deploying the virus wholesale as an insecticide would be to advance the time at which the insects are exposed to it. In nature, viruses often kill off insects late in their life cycle, long after the pests have done their damage. With earlier infection, the caterpillars would die correspondingly younger, and would thus release a smaller number of virus particles.

Rifle rather than shotgun

The next decade will undoubtedly see a tremendous increase in research and development devoted to micro-bial insecticides. At present, worldwide investment in industrial investigations into methods of controlling plant pests with microbes is thought to total no more than 600,000 dollars per year. That contrasts with the 100 million dollars spent by industry on the development of chemical pesticides. But the balance has already shifted, as the benefits of the biological approach, and the draw-backs of chemical warfare, have become more apparent. Not the least compelling feature of biological methods from the financial standpoint is that the benefits of a pest eradication scheme may persist for much longer than is at first envisaged. The cost of introducing the European spruce sawfly virus into Canada in 1959 was probably less than 50,000 Canadian dollars, but the campaign at that time has had lasting effects, and seems to have contributed to the control of the fly ever since.

Pest control by microbes also looks increasingly attrac-tive to the agriculturalist and ecologist. A vividly clear lesson from recent decades is that of the heavy price paid for the appalling crudity with which we have sought to 'have dominion over nature'. Rachel Carson's *Silent Spring* was the book that first propelled the subject into public prominence, and the most grotesque offence she high-lighted was the indiscriminate spraying and deployment of pesticides. We believed we could ignore the effects of such practices on the rich profusion of life so long as the target species was demolished — like a doctor who hands out broad-spectrum antibiotics for trivial infections, to attack microbes he has not stopped to identify (or even those whose existence he has not bothered to confirm), or a policeman who wields a shotgun to shoot a suspect murderer in a crowded theatre, in the dark.

Microbial insecticides should allow us to avoid the unhappy repercussions of that sort of behaviour. They are innately safe and have a restricted range of activity. A single, tailor-made insecticide can be used to attack a single species of insect pest. But this does not mean that, having narrowed our aim, put on the light and taken a rifle off the shelf in place of the shotgun, we can afford to forget about the secondary, tertiary and long-term consequences of our actions on other species of life. The second commendable feature of biological control, indeed, is that by its very nature it makes us think of the entire ecological community with which we seek to interfere. This is also true of 'integrated control', using both microbes and judiciously applied chemicals. Though the dangers may be less than those created by the maniac wielding toxic chemicals by the lorry-load, our eventual success will depend critically on the ecological sensitivity and foresight with which we harness the undoubted potency of our microbial allies.

5 For the Gourmet

It may seem imprudent, in a book arguing that we should find common cause with our tiniest allies in tackling profound and harrowing global problems, to devote a large part of even one chapter to the subject of booze. That is a risk I am delighted to take. For the truth is that the fermentation of wine, beer and other alcoholic beverages is one of the most venerable and universal of man's domestic activities. Another is the leavening of bread by yeast. Considered together, and alongside the production of countless other foods and drinks — from cheese and yoghurt to pickles and sauerkraut — the making of bread, wine and beer represents a massive microbial contribution to human nutrition and gastronomy. As well as having limitless potential as novel foods of the future (chapter 6), micro-organisms are crucial in producing our food and drink today.

The two main categories of fermented drinks, beer and wine, differ in both raw materials and methods of preparation. But microbes, principally yeasts, are the executive agents in fashioning both splendid varieties. Wine is made from the juice of grapes (and to some extent from such fruits as elderberries, bananas and apricots). Beer is based on grain, usually barley. In each case, though scientific knowledge plays a greater part than in the past, traditional craft, the accumulation of centuries of human kinship with the microbial craftsmen, is paramount. 'A mystery *and* a science' is how R. H. Courage, chairman of Courage Barclay and Simonds, has described brewing to readers of *The Times*. If this sounds like an attempt to have it both ways, to claim the resources of both the Vatican and the National Physical Laboratory, it is nonetheless an assertion with which all dedicated brewers would agree. Every good *maître de chai* would feel the same way about

his craft. Science is useful, but the essence of real wine-
and beer-making fortunately remains beyond quantitative analysis.

Yeasts of the vine

Archaeological evidence suggests that man was making
wine as early as Mesolithic times, over ten thousand years
ago. The art has certainly been practised in Egypt for over
two thousand years. It was the Greeks, before the Christian
era, who first created vineyards in southern Italy, France
and other parts of southern Europe. This spread of wine-
making, encouraged by the excellent climate, was so successful that eventually the Egyptians ran down their own
industry and began importing wine from Greece. The
Romans took the organization of wine-making a huge
and highly efficient step further, to cater for the needs of
their legionaries. It was they who established the vineyards of Alsace and the valley of the Mosel.

Though its underlying microbial magnificence passeth
understanding, the elements of wine-making are simply
stated. After gathering, the grapes are crushed to provide
the juice or *must*. Being both acidic and rich in sugars
(mainly glucose and fructose), must is an ideal medium
for the growth of yeasts. For red wine, black grapes are
used, including skins and stalks. White wine is usually
made from either black or white grapes, without their
skins. American wine bibbers, however, do not appreciate
the subtle components of flavour given by the stalks, so
these are always excluded in US wine-making.

The waxy 'bloom' on the surface of grapes contains
many different microbes, including the wine yeast *Saccharomyces cerevisiae* var. *ellipsoideus*. The strains of yeast
are characteristic of both the grape and the district. At
least 150 are known. The must may be transferred without
further treatment, into a fermentation vat (usually made
of wood or stone and holding anything from 250 to 200,000
litres or more). The resident yeasts then promote a vigorous fermentation, turning the sugar into alcohol and

triggering the initial magical chemistry that leads eventually to the characteristic flavour and aroma of the wine.
Increasingly, however, 'desirable' strains of yeast are
admitted to initiate fermentation. Particularly in the US,
modern technocratic wine manufacturers now tend to go
one stage further. They destroy the natural microbes
totally by pasteurizing the must or by bubbling sulphur
dioxide gas through it. Then they introduce a standard,
defined strain of yeast. This avoids the occasional uncer-
tainties of traditional methods. Whether it improves the
quality of the final product is another matter. Sulphuring
certainly ruins the flavour of some otherwise passable
plonk. Cheap white wine in particular often reeks with
the sulphurous acidity of sulphur dioxide.

Without scientific attention, natural fermentation process is remarkably efficient. There is a natural succession
of activity in the vat, in which the early yeasts that initiate
fermentation die away after producing four per cent or so
of alcohol. They are followed by the wine yeast itself,
which converts more sugar into alcohol, up to a concentration of twelve per cent or so. (Fortified wines, such as
port, sherry and Madeira, have extra alcohol added to
them later.)

During fermentation, the must in the vat becomes
warm and bubbles with carbon dioxide — the other
principal product from the prodigious work of the yeasts
in fermenting the grape sugars. The *vigneron* usually
allows fermentation to proceed to completion — until
the yeasts have used up most of the sugar or are inhibited
by the amount of alcohol they have generated. This can
take from a few days up to a couple of weeks. Alternatively,
the scientifically minded wine-maker can cut the process
short by pasteurization or by sulphuring.

The young wine next goes into wooden casks in a cool
cellar for ageing. During this period, and later after
bottling, further changes crucial to the final flavour and
aroma of the wine occur. As the yeast cells and other sediment settle out, the wine is racked into fresh casks from

time to time, over several months. Finally, the wine is 'fined' with isinglass to clear the last traces of turbidity. (*Vin ordinaire* is merely filtered.)

There are many variations in the techniques used in making different wines. Rosé, for example, is often made by pressing black grapes only lightly, and following this with a relatively speedy fermentation (at a higher temperature than usual) before the skins are removed. Otherwise, it is produced from grapes with light-rëd skins, and sometimes (though connoisseurs frown upon this practice) by mixing together red and white grapes.

Genuine sparkling wines, like champagne, sparkle because of a second fermentation that has taken place inside the bottle. (Non-genuine types are pumped full of carbon dioxide, like lemonade.) After fermentation and storing in a cask, young champagne is bottled and sugar and yeast are added. The yeast produces more alcohol and carbon dioxide — which cannot escape from the tightly corked bottle and thus remains dissolved. In authentic champagne-making, a skilled *dégorgeur* removes the dead yeast cells in the tricky manipulation of *dégorgement*. The bottles lie on their sides at first, on a table that can be tilted. Each day, the *dégorgeur* twists each one slightly to dislodge the sediment, tilting it a little more until eventually the bottles are standing on their necks. Then, after a week, comes the finale, as he releases each cork very briefly, just enough to allow a tiny amount of wine, containing sediment, to escape. Sometimes the neck of the bottle is frozen, so that the yeast comes away as a solid plug. Momentarily stopping the bottle with his thumb, the *dégorgeur* then inserts and wires a new cork.

Another variation is the solera system employed in sherry-making. The solera is a tier of five or more casks, the top one of which receives the wine after fermentation and racking. Over a period of five or six years, the wine is transferred successively from cask to cask, and eventually drawn off from the bottom one to be fined, sweetened, fortified with brandy and blended. Inside the casks, a *flor* yeast grows on the surface of the wine, and its meticulous, unhurried activity is responsible for the eventual fine flavour of the sherry. The process is one that no micro-biological or chemical engineer would ever devise. Yet, by providing perfect conditions for the aristocratic *flor* yeast to operate, it works superbly well.

Although alcohol is the principal chemical product of wine-making — and one whose minimum concentration is usually defined in law — it is the countless other fruits of microbial industry, many of them present in vanishing-ly small quantities, that determine the quality of the wine. Far beyond analysis or even rough description, the forma-tion of these constituents during fermentation, and their subtle alteration and interaction during ageing, accounts for both the contrasts between wines from different regions and the variations in quality from the same vine-yard. A poor growing season, which yields grapes con-taining too little sugar and too much acid — or indeed much less obvious adversities — can mean that the product of a noble château has to be despatched to the distillers for making inferior brandy.

Oenologists tend to become livid when they hear ana-lytical chemists conferring about wine. Attempts to define in chemical terms the character and composition of fine wines are both sacrilegious and destined to abject failure. Certainly, such efforts have met with little success to date. Biochemists claim that chemicals called esters, formed during fermentation and maturation — we know some of them as flavouring agents in cheap sweets, such as 'pear drops' (amyl acetate) — have a potent influence on bou-quet. But the crudity of this type of analysis has been matched so far only by the utter failure of attempts (partic-ularly in the US) to accelerate the ageing of wine. Espe-cially with red wines, this process can take many years. In all probability, we shall never come near to discerning the subtle molecular modifications involved.

Of all microbial craftsmen, the wine yeasts are the most magnificent. Industrious aristocrats, they generate vast quantities of wine – currently some sixty-five million hectolitres each year in the world leaders, France and Italy. They are also responsible for the sublime clarets of the Medoc, and for the fine hocks of the Rhinegau. (And that, in turn, means that as well as being highly prized in themselves, excellent wines are excellent investments.)

Perhaps the final word, for those who love good wine, should be left with C. L. Duddington, writing on the American approach:

The regularity of the weather and the great care taken in American wineries to see that all processes are scientifically controlled as far as is possible result in a uniform product. The alcohol content of California wines is high; one of their burgundies has been recorded as containing nearly 15 per cent of alcohol, a figure that could not be attained in Europe.

The magnificent microbes are capable of far greater attainments than that – and more meaningful ones.

Brewing, ancient and modern

To the brewer and tippler, brewing is every much as revered an art as wine-making. It is probably almost as venerable too. Beer seems to have been brewed in Mesopotamia at least six thousand years ago. Ancient Egypt apparently had a well developed brewing industry – one that is thought to have been nationalized. Though brewing never became established in the vine-growing countries on the Mediterranean, it was again the Greeks who brought the art to western Europe. The Romans introduced 'modern' brewing of malted barley to Britain (where some beer had been produced previously from malted wheat). During the Middle Ages in Britain, the monasteries were the centres of brewing. Although many of the earliest brews resembled today's 'barley wine', it seems that drinkers had a range of types of beer – probably as wide as that available in the average pub today. Some 35,000,000 barrels of beer are now brewed each year in Britain.

The chief way in which brewing differs from wine-making is in the need for an extra stage at the very beginning. The insoluble starch in the barley grains has to be broken down to simpler, soluble substances, principally maltose (malt sugar), which yeast can ferment to yield alcohol. This 'saccharification' begins during *malting*. After being steeped in water, the barley is spread on the floor, or placed in slowly revolving drums, for about a week. The grain starts to germinate during this time, producing a young shoot and root. Enzymes capable of attacking the starch are formed too, and they begin their work.

At a point determined by experience and tradition, the maltster arrests germination by kilning, raising the temperature to about 80 degrees centigrade. This leaves the malt in a condition in which it can be stored until needed, but destroys little of the enzyme activity in the grain. Malt is heated to a higher temperature for making darker beers such as stout. Though barley malt is by far the commonest sort, other grains are also used for making some beverages. They include oats for oatmeal stout and rye for the Russian *kvass*.

Brewing proper begins with mashing. The *grist* — a mixture of malts, crushed to allow enzymes access to the starch; a little unmalted grain and some sugar — is mixed with water and warmed in a mash tin. The malt enzymes now resume their work, releasing maltose and other substances. After mashing, the resulting wort is drawn off, leaving behind a solid sediment. Sparging, the percolation of water through the mash, extracts more sugars. Mashing techniques vary. In Britain brewers mix the mash with warm water at about 65 degrees centigrade and leave it for two hours. In the European and American system, used in making lager-type beers, the mash is heated up in stages.

Next, the wort is boiled. This both arrests the enzyme action and sterilizes the wort of all microbes. The most important feature at this stage, however, is the addition of hops — dried female flowers of the hop plant. Tannins from the hops help to precipitate unwanted protein as an insoluble sludge, while other substances contribute to the beer's characteristic bitter flavour.

The crucial fermentation follows. With even greater fastidiousness than their colleagues making wine, brewers have nurtured, cultured and preserved pedigree yeasts for this aristocratic task. The pioneer of strain selection was Emil Christian Hansen, a Danish contemporary of Pasteur who worked at the Carlsberg brewery in Copenhagen. He found that the precise strain of yeast used greatly influences the type and quality of beer produced.

Countless special brewing yeasts have now been isolated and jealously preserved throughout the world. Yet they all fall into two fundamental groups. First are the top yeasts, so called because they rise to the surface of the beer during fermentation. Vigorous fermenters, they grow best at a relatively high temperature (20 degrees centigrade) and brewers use them to make beers strong in alcohol, such as the English ales. Apart from Britain, top fermentation is practised widely in Belgium and Canada.

Most top fermentation beers are brewed by what is called the skimming system. After cooling, the wort passes into a 4,000-10,000 litre vat, where it is 'pitched' (inoculated) with a strain of *Saccharomyces cerevisiae* particularly cherished by that brewery. As in wine-making, a vigorous fermentation soon sets in, as the yeast works on the sugars to produce alcohol and carbon dioxide. From time to time the brewer skims off the head of excess yeast, which floats to the surface. He must also ensure that the temperature does not rise too far. Then, after fermentation is complete — within a few days for proper beer, or a matter of hours for ultra-modern varieties — he racks off the beer into barrels or storage tanks. After fining and storage at the brewery for a week or more, during which some secondary fermentation may occur, the beer is ready for the customer.

It is bottom yeasts, however, that take credit for three-quarters of the world's beer output. Working in the depths of the vat, they ferment slowly, function most effectively at a temperature of 12-15 degrees centigrade, and produce light beers with low alcoholic content. Bottom fermentation probably originated in Bavaria. One reason that it is popular in Central Europe and America is that beers produced by this process are stable despite higher summer temperatures. Conversely, unlike strong ales, these light beers do not become turbid when kept in a refrigerator.

Saccharomyces carlsbergensis is the yeast that provides us with lager and similar beers. After a fermentation lasting a week or more, the beer is 'lagered' — kept in huge tanks at 0-3 degrees centigrade for several weeks or months. During this time the yeast promotes a slow secondary fermentation whose products engender the characteristic mellow flavour. The beer is then filtered, and perhaps pasteurized.

Much has been written in recent years about the demise of decent beer, as the range of beers available diminishes under the drive for scientific standardization. From our present standpoint, one cannot but lament the loss of the diversity of human and microbial skills, symbiotic craftsmanship built up over many centuries, as the technological admass takes over.

Compare, for example, beers made by the great Burton Union system, and those now being fabricated via the physically and chemically precise process of 'continuous culture'. In the Burton system, part-fermented wort passes into a series of casks for further fermentation. Each cask is topped by a 'swan neck' pipe, the end of which bends over into a long trough. Propelled by carbon dioxide, yeast rises up the swan necks and into the trough, which slopes towards one end. It then passes down to the end of the trough, where it enters another tube, which returns it to the casks through their bung holes. As with the solera, no technocrat would have devised such an arrangement. It makes little obvious scientific sense. What it does do is to provide the conditions under which the yeast can produce that superb and characteristic flavour of Burton beer.

What the chemical and microbiological engineer between them do delight in is the continuous culture process. Like nitric acid or polythene manufacture, raw materials are metered in continuously, all conditions of temperature, acidity etc. are rigorously stabilized, and the product emerges as a continuous stream of effluent at the other end. Some brewers have been persuaded to try this with beer. But the resulting material comes nowhere near the quality of real beer.

Diversity of liquor

Turning quickly away from such misplaced efforts, an area where human ingenuity is genuinely impressive is in the countless different stratagems devised throughout history to make alcoholic liquors. Many Indians in Central and South America, for example, have discovered a neat source of the starch-splitting enzymes needed to help yeasts and other microbes attack corn grains: their own saliva. By chewing the grains for a while, and then spitting the mixture of corn and saliva into a bowl, they set the stage for a natural fermentation promoted by microbes in the environment.

More sophisticated is the technique used to make Japanese *saké*, in which microbes engineer both the release of sugar from rice starch, and its fermentation to produce alcohol. The starting material is boiled rice, to which a mould, *Aspergillus oryzae*, is added. The mould forms enzymes that break down the starch to sugar, which the yeast *Saccharomyces saké* in turn converts to alcohol (up to twenty-two per cent). The microbes also produce some lactic acid, which contributes to the flavour of the *saké*.

Cider-making is one of the most natural processes still used to make an alcoholic beverage now manufactured in commercial quantities. So far, science and standardization have had little sway here. Indeed, it is only comparatively recently that the factory has replaced the farm as the centre of cider-making. Cider apples, of which there are several classical varieties, differ from both cooking and eating apples in having a high content of tannins, which give them an astringent flavour. They are gathered in the autumn, and pulped to yield the juice, which is then poured into fermentation vessels fitted with valves that allow carbon dioxide to escape but will not admit air. This ensures anaerobic conditions.

Microbes from the surface of the apples — yeasts, bacteria and moulds — conduct the spontaneous fermentation, which lasts for about four to six weeks, during which time the organisms convert fructose and other sugars to alcohol. The cider-maker chooses with care the moment to stop fermentation. He may either arrest the process early, leaving enough sugar to provide a naturally sweet cider, or wait until completion to give a drier product or one that can be sweetened by adding sugar if desired. After filtering the cider, he allows it to mature for some weeks or months, before casking or bottling it. Even here, science has begun to make inroads. Particularly in Europe, 'dominant fermentation' has become popular with some cider-makers who prefer to demolish the natural microbial flora of the apples with a massive dose of sulphur dioxide, and then add an equally huge slug of pure yeast.

Whisky and the other spirits
Strictly speaking, the production of whisky and other spirits can scarcely be considered as microbiological craftsmanship. Its best known stage, distillation, is indeed a purely physical process, one that chemists use to purify substances in the laboratory. The basis of it is simply that when two liquids, such as water and alcohol, with different boiling points are mixed together, the more volatile one can be concentrated by heating the mixture and condensing the vapour. But distillation is only the most dramatic

flourish in making spirits. The starting materials — chiefly beers and wines — are of course microbiological products. And the point of distillation is not merely to manufacture absolute alcohol. It is to create a beverage, whether whisky or brandy, whose final excellence owes much to the complex and mysterious fermentation processes earlier.

Two basic types of stills are employed in making spirits. The simplest and oldest is the pot still, which Scotsmen use for their fine malt whiskies, and which the French use, in a slightly modified form, to make brandy. A large round metal container, the pot still sits in a furnace, with its tapered head conducting the vapours away to a cooled condenser. Pot stills have two disadvantages, one of which is not a real disadvantage at all. They are not very efficient in separating water and alcohol — certainly far less so than those employed by chemists to purify chemicals with great precision. But we do not only wish to produce pure alcohol: we desire the glorious fruits of microbial alchemy. A somewhat inefficient piece of apparatus is ideal for that purpose.

The other shortcoming of pot stills is more significant. They distil one batch of spirits at a time, and need successive reloading. Being a physical rather than a vital living process, however, it is here permissible to introduce the continuous processing principle. This has been done in the Coffey still, named after the Irishman Aeneas Coffey, who invented it. This type of still works without respite, and yields a more predictable, standardized product. Alas, the price of this greater efficiency is that the Coffey still removes too much and too many of what to the chemist are impurities, but what to the whisky connoisseur are ingredients of heavenly nectar.

The earlier stages of whisky-making are not unlike brewing. For real Scotch — malt whisky — the wort is made entirely from malted barley. For ordinary Scotch, some unmalted grain is included. Irish distillers normally used a mixed grist of barley, oats, wheat and rye.

Bourbon comes from a mash consisting mainly of corn (maize), while American rye whisky is made from an assortment of rye grains. Whatever the grain, it is crushed and mixed with hot water, whereupon enzymes begin to convert the starch into maltose.

After transferring the wort to a fermenting vat, the distiller adds *Saccharomyces cerevisiae* to ferment the sugar in the wort. Bacteria also play an important role, producing lactic acid and other substances. After two or three days the fermented wort (no hops are added) is transferred to the still.

Malt whisky comes chiefly from the Highlands, from the Glenlivet and Speyside districts, and it bears traces of the unique water of those areas and of the peat fires used to dry the malt. The Lowland malts, principally Islay and Campbeltown, are less smooth. Malt whisky is made in pot stills, so retaining the rich variety of its microbial endowment. Afterwards, many years of maturation in the cask are needed to bring out the full aroma and bouquet.

Ordinary Scotch is made in continuous stills, mostly around Glasgow. Though a more uniform product, it has none of the splendour of the malts. As well as being sold as a routine drink, it is much used in blending. Indeed, totally unblended whisky is uncommon — particularly outside Scotland.

Next to whisky, brandy is the finest of the spirits. The term covers only spirits produced by distilling grape wine: it excludes those made from wine based on other fruits, such as kirsch from cherries. The greatest and best known brandy is, of course, French cognac, from the *départements* of the Charente and Charente Maritime, closely followed by armagnac from the *département* of Gers. Considerable quantities are also produced for fortifying wine. In making brandy, the grapes are first fermented to a white wine which, without sulphuring but with pre-warming, is run straight into pot stills. The early and late distillate is rejected. The middle fraction is combined with those from other batches and redistilled. Again, the middle distillate is taken and then matured in oak casks for anything up to twenty years.

Then there is gin – little more than dilute alcohol flavoured with herbs. Fermentation of maize or rye wort yields the initial spirit, which acquires its characteristic taste by being redistilled together with juniper berries and other 'botanicals'. Rum is the end-product after distilling fermented juice of sugar molasses, usually in continuous stills.

These and countless other alcoholic drinks, from the roughest firewater to the smoothest and most noble malt whisky, from the simplest plonk to the greatest claret, are all products of the industry and versatility of the microbes. Whether we appreciate the elegance of Château Lafite, or merely wish to get drunk as quickly as possible, we have them to thank for the liquid in the bottle. The taxman should appreciate their efforts too. Britain's Customs and Excise Commissioners received well over £ 1000 million in duty on alcoholic drink in the financial year 1974-75.

The baker and his yeast

Bread-baking, which dates back to the Stone Age, is almost as universal as the fermenting of alcoholic drinks. It too depends upon microbial action. Though the need here is not for alcohol but for ample carbon dioxide gas to raise the bread, the strains of *Saccharomyces cerevesiae* used nowadays came originally from the brewery. To leaven his dough, the baker simply mixes in some yeast, and stands it in a warm place for a few hours. He then pops the dough into the oven – or, in a baking factory, on to the continuously moving conveyor belt passing through the oven.

Decades ago, there were sufficient starch-splitting enzymes in flour to release sugar for the yeast to work on. Nowadays, sugar has to be added, because enzymes are among the many constituents removed during 'purification' of white flour. The time between adding the yeast and the start of baking is critical. Further fermentation occurs in the oven, yielding carbon dioxide that determines the bread's fluffy texture, until the heat inactivates the yeast. The heat also drives off most of the alcohol. Nonetheless, a new loaf can contain as much as 0.5 per cent of alcohol — about a sixth as much as in mild ale.

It might appear that chemicals could replace microbes in the process of leavening. So it seemed to the great German chemist Baron Justus von Liebig who, over a century ago, invented baking powder — a mixture of chemicals that form carbon dioxide when moistened. But Liebig was disappointed. Baking powder produces highly unsatisfactory bread, and has proved useless for that purpose. Clearly, carbon dioxide is only part of the story. The yeast encourages other, more subtle, physical and chemical changes in the dough, of which we are yet quite ignorant.

Almost up to the beginning of this century, bakers allowed dough to leaven spontaneously. They simply retained until next time a little of the successful dough from one baking, which had produced satisfactory loaves of the right flavour and consistency. Thus the microbes, which included not only yeasts but bacteria too, would be conserved in the sour dough or 'sauerteig' of the bakery. Modern sanitary ideas, and the need for rigorous standardization of the product, have brought the virtual eclipse of this system.

However, bacteria that produce lactic acid and thus mild souring are required in making rye bread and some other breads. They can be added as pure cultures, but a few bakeries still use the sauerteig process for the purpose. Most recently, bakers and microbiologists, putting their heads together, have begun to realize just how much we have lost in the flavour of bread by rigorously excluding bacteria. Experiments have thus been in progress to initiate the sauerteig procedure, in modern guise, by adding favoured bacteria to the baker's yeast. No doubt techniques of this sort, based on an age-old domestic use of microbial skill, will shortly be announced as new research discoveries.

Glorious cheese

Milk goes sour when bacteria break down the milk sugar (lactose), producing lactic acid which in turn curdles the casein and other proteins. It is in effect a method of preserving what would otherwise be an unstable food, because the acid stops destructive microbes from growing. When early man first learned, presumably by accident, to make cheese and other fermented milk products, these foods were probably valued initially for this reason. Today, of course, the great cheeses of the world are among our most prized gastronomic delicacies.

The first step in cheese-making is the curdling of the milk proteins, giving a solid mass from which much of the water is drained away. Coagulation may be microbial, as in soured milk. Or it can be triggered off with rennin, an enzyme formerly extracted from the stomach of calves but now increasingly made microbiologically by the mould *Mucor miehei*. Cottage-type or cream cheeses are made simply by adding bacteria such as *Leuconostoc* to pasteurized milk. Lactic acid formed by the bacteria precipitates the curd, which is cut into cubes, and made firm by gently heating. Salt and perhaps a little cream are added before packaging.

Most cheeses, however, and certainly the best ones, have to be ripened by bacteria and fungi. Though initially almost all natural cheeses look alike, differences in the attendant microbes and the conditions under which they work create the wide diversity of cheeses as we see them on the cheese board. With Cheddar, for example, the same bacteria that (together with rennin) produce the curd also ripen the cheese. As they die, their cells release enzymes which, working on the milk fat and proteins, form the many different compounds that give Cheddar its characteristic flavour. The nutritive value of the cheese also rises greatly, as the bacteria synthesize vitamins, particularly those of the B complex.

The initial coagulation occurs rapidly (in about twenty to forty minutes) and the curd is then cheddared — heated,

cut into pieces and stacked to force out the whey. The
ripening of Cheddar, after salting and packing into hoops
lined with cheesecloth, can take several months. In the
early weeks of ripening, the number of bacteria reaches
hundreds of millions per gram.

Other hard curd cheeses include the venerable Roque-
fort, which is made from the milk of ewes and is ripened
by the fungus *Penicillium roqueforti*, and Stilton and
Gorgonzola — in both of which this fungus is the principal
ripener and the source of the characteristic blue veins.
Propionic acid bacteria promote the ripening of Swiss
cheese. They turn the lactic acid in the curd into propionic
and other acids (which give the well-known flavour) and
carbon dioxide — which produces the eyes or holes in the
cheese. (In the manufacture of 'processed' cheeses of the
sort that are particularly popular in the United States,
the same microbes are considered to be unwelcome agents
of decay.)

The soft and semi-soft cheeses — Camembert, Limberg-
er and others — owe their consistency and flavour to microbes that soften the curd during ripening. A cosmo-
politan population of bacteria, moulds and yeasts lives
on the surface of the cheese in a slime that contains as
many as ten billion microbes per gram. Their enzymes
diffuse into the cheese, softening it and creating its characteristic taste and aroma. The chief organism in the
ripening of Camembert is *Penicillium camemberti*. Many
others are not only tolerated but welcome — the outer
skin of Camembert cheese contains a massive number and
assortment of microbes. Cheese-makers usually inoculate
new batches with a surface smear from an older cheese,
but do not seek to suppress other organisms.

As in brewing, having the right starter organisms — for
both stages of cheese-making — is crucial to success. But
so far at least, the drive for simplicity and mechanization
seen in the brewing industry has not encroached into
cheese-making. One attempt in this direction met with
comic failure. About twenty years ago, a group of scientists

succeeded in making something which, in appearance
and chemical composition, was identical with Cheddar
cheese. They started with sterile milk, devoid of all micro-
organisms, and used a chemical (gluconic acid lactone)
to precipitate the curd. But their work was soon justly
forgotten, consigned to the annals of misplaced science.
The Cheddar substitute they produced had no cheese
flavour whatever.

Butter, and microbial milks

Butter is a microbiological product too — the cream has to
curdle if the butterfat is to separate satisfactorily during
churning. Dairymen usually add two species of bacteria
to the cream — *Leuconostoc citrovorum*, which helps create
the characteristic flavour, and *Streptococcus cremoris*, which
forms the lactic acid that causes souring. *Leuconostoc* produces a substance called diacetyl, which gives butter its
taste and aroma. Neither organism by itself yields good-
quality butter.

The bacteria that take such a mammoth part in cheese-
making, lactobacilli, also play the key role in the manufacture of various fermented milk beverages. *Kefir*, which
originated in the mountains of the Caucasus, is made by
inoculating 'kefir grains' into the milk of various domestic
animals. The grains are small, cauliflower-like granules
containing dried masses of lactobacilli and yeasts. When
added to milk in goatskin bags, they release their microbes,
which promote an intriguing form of fermentation. The
bacteria form lactic acid and the yeasts produce alcohol
and carbon dioxide. The gas cannot escape, and the result
is a unique, effervescent, alcoholic buttermilk.

Yoghurt, originally a fermented milk made by Bulga-
rian tribesmen but now found in every Western supermarket and delicatessen, is a similar product, though
lacking alcohol. *Lactobacillus bulgaricus* ferments the milk,
producing the uncomplicated flavour of lactic acid. Among
microbiological milk delights not yet on our supermarket
shelves are the *busa* of Turkestan, the *koumiss* of Central

Asia and the *leben* of Egypt. American consumers, however, have the pleasure of consuming buttermilk, which
is milk (usually skimmed) soured by bacteria and then
beaten up to produce a smooth creamy beverage.

Especially interesting for its medical importance is
acidophilus milk. This is made by sterilizing whole milk,
inoculating it with *Lactobacillus acidophilus* and allowing
the bacteria to flourish until they have created the required degree of acidity. It has proven value in treating intestinal conditions caused by an imbalance of putrefactive
bacteria in the lower alimentary tract. When consumed,
bacteria in the milk pass through the stomach and into the
lower gut in sufficient numbers to establish themselves
there, replacing the offensive organisms, which are driven
out by the acid produced by the lactobacilli.

Sauerkraut, pickles, and silage

Lactobacilli are also the principal creators of fermented
foods like sauerkraut and pickles. The bacteria originate
from the plants used as raw materials and they ferment
the sugars in the plant cells to give lactic and other acids
which both flavour the foods and help preserve them.
Ensilage of cattle fodder rests on the same principle. Once
plant tissues have been turned into silage, they can be kept
indefinitely without risk of decomposition.

Sauerkraut is fermented cabbage. The cabbage is
shredded and salt added. It is then either left to ferment
naturally or inoculated with lactobacilli or old sauerkraut
as a starter. The salt brings out the cabbage juices and
encourages the right bacteria to proliferate. In commercial
production, the fermentation proceeds in large vats.
Wooden frames keep the cabbage submerged and thus
maintain anaerobic conditions. Fermentation, which takes
two to three months, is a sequential process in which various bacteria play their distinctive parts in creating the
eventual complex and savoury chemistry. Acids, esters
and diacetyl (the characteristic flavouring agent of butter)
all contribute to the aroma.

Fermentation by a natural microbial flora also provides us with pickles and related delicacies. Here a high concentration of salt, which inhibits many microbes, ensures an even more selective fermentation than with sauerkraut. Cucumbers, usually the main constituent of pickles, are harvested while small and unripe, and are 'cured' over two months or so in a brine vat. Again, there is complicated microbial activity during this time, involving both bacteria and yeasts. Lactic and other acids appear, and the consistency and flavour of the cucumbers change. The pickles are then transferred to other tanks with fresh water and heated, to remove the excess salt. Packed in a fluid containing vinegar, sugar, spices and other flavourings, they are then ready for the table. Weaker brine is used in curing dill pickles, and the sugar and vinegar are often added at the beginning of fermentation.

Of the many oriental foods produced microbiologically, Nam-pla, a fermented fish particularly popular in Thailand, is one of the most exquisite. It is made from several species of small fish which, mixed with salt, are fermented by a varied population of bacteria in sealed tanks for about six months. The final product is a succulent dark brown liquid, whose composition has so far defied chemical analysis. Though having a very different taste, its appearance resembles that of soya sauce — another microbial product. Soya sauce is based on soya bean flour and wheat bran, fermented by moulds and yeasts.

Leaving for the moment our own gastronomic indebtedness to the microbes, this is an appropriate moment to record one particular contribution they make to animal nutrition. Silage, a product of the labours of innumerable bacteria including *Lactobacillus bulgaricus*, results from a process similar to those used to make pickles and sauerkraut. For the cattle that eat it, silage is doubtless every bit as succulent.

Like cheese and several other microbiological foods, silage was originally devised as a preserved form of food — an alternative to hay. Farmers can make excellent silage even in a rainy season when hay-making is impossible. This is a traditional example of a key microbial skill with great importance in the modern world: the production of food independently of adverse weather conditions.

Some farmers make silage from grass, corn stalks, alfalfa or maize. Others grow a mixed crop especially suited to ensiling, often a blend of oats, beans and vetches. The plant material is chopped and then tightly packed, either into a tall cylindrical silo or, sometimes in Britain, into a pit. Either way, compression is important to ensure a favourable anaerobic environment for the bacteria to perform. If the crop contains relatively little fermentable sugar, the farmer may add some molasses, to provide sugar for the bacteria to start to work on.

The microbes soon begin to proliferate luxuriantly in the plant juices and to ferment the carbohydrates, causing the silage to become warm. Bacterial activity yields lactic and other acids and also diacetyl, the flavour of which is particularly relished by cattle. After three or four weeks, the process slows down and the temperature falls. The silage is then ready for eating, or for limitless storage. It shows the original structure of the plant material that went into it, but in a well-preserved state, sweet-smelling and luscious to animals.

Scientists have tried to better the microbes here too. In principle, imitating natural ensiling should be easy. The critical microbial products are acids, which preserve the silage, so why not simply dowse grass in acid? This simplistic stratagem is, in fact, incorporated in the AIV silage process, which is used in Scandinavia. The farmer just puts his crop into a silo and pours hydrochloric and sulphuric acids on top — which ensures that no microbes of any description can live there. In theory, AIV silage is wholesome, provided that the farmer mixes chalk with it to neutralize the acidity before feeding it to his livestock. But most farmers believe that the microbial version is both safer and nutritionally superior. The process has never caught on to any significant extent outside Scandinavia. Chemical silage is not popular with cattle.

Vinegar and its brown imitations

Nowhere is the gulf between science and microbial crafts-manship more apparent than in vinegar-making. It can be appreciated at first hand by successively tasting authen-tic vinegar, made by traditional fermentation, and the 'vinegar type condiment' that masquerades under the same name. Chemical analysis tells us that acetic acid is the principal component of vinegar. Hence (the techno-crats argue) we can replace vinegar by a solution of acetic acid — suitably coloured, because we have become used to the idea that vinegar is brown. But this is as great a fallacy as supposing one can substitute a fine malt whisky by a solution of the same concentration of alcohol. Without the superb micro-constituents of genuine vinegar, fashioned microbiologically, the synthetic mimic is meaningless. Those who believe otherwise can be compared only with an impresario who sacks a symphony orchestra and takes on a noisy organist in its place.

Acetic acid results from the oxidation of alcohol, and the traditional method of making vinegar is by spontaneous souring of wine. The word itself derives from the French *vinaigre*, which means sour wine, though today cider and other forms of alcohol are also used as starting materials. In the traditional Orleans process, which is still employed in France, wooden vats or casks are partially filled with wine, which becomes colonized by *Acetobacter* and other bacteria capable of oxidizing the alcohol and creating the other flavouring constituents. The organisms grow on the surface of the liquid, and the oxidation takes many weeks, being limited by the rate at which oxygen from the air can diffuse into the wine. But the Orleans process survives because of the superb quality of the vinegar it produces.

A more rapid technique, widely used today, was devised

in Germany during the early nineteenth century. Wine or other liquor trickles in a continuous stream through a tank loosely filled with wooden shavings or similar filler material. Air percolates in the opposite direction, and the acetic acid bacteria grow vigorously on the surface of the shavings. Different countries favour different starting fluids. Most vinegar in England comes from malt, while cider vinegar is popular in the US, and wine vinegar predominates on the Continent. The Scots have an understandable liking for distilled malt vinegar. As with alcoholic beverages, the final flavour and aroma is distinctive of both the starting materials and the ensuing microbial activity. Acetic acid is merely the chief constituent. The full, rich flavour is determined by the esters, sugars, volatile oils and the myriad other substances which the bacteria produce.

Dependent eaters and imbibers
It is remotely possible that some of those reading this book are teetotallers who never eat bread, cheese, yoghurt, pickles, sauerkraut or other microbiological foods, and never put vinegar on their fish or use mayonnaise or salad dressing. They may feel less conscious of microbial munificence than the rest of us. Yet they have probably drunk cocoa, the flavour of which owes much to a microbial fermentation that helps remove the beans from the pulp covering them in the pod. They will almost certainly have enjoyed coffee or black tea, both of which are thought to derive part of their flavour through bacterial action — during the fermentation of tea and of the coffee berries when the beans are soaked in water to loosen the berry skins before roasting. And they may well have relished the flavours introduced by microbes during the curing of ham and other meats.

No individual eater or imbiber, no cuisine, is independent of the busy and ubiquitous microbe. Those microbial contributions to nutrition and gastronomy described in this chapter are the principal ones, and the most nearly universal. There are many more, often of great subtlety, and doubtless others as yet unrecognized. As in agriculture, we find again that the microbes are global benefactors. Yet when we have recognized their crucial roles in producing our food and drink, and have tried to replace them by technical ingenuity, our efforts have ranged from the inept to the abysmal.

6 Foods and Feeds of the Future

We know more today than ever before about the science and art of agriculture. Enormous increases in output have followed improvements in farm mechanization, pest control, water management, plant and animal breeding and soil fertilization. Our knowledge of human and animal nutrition makes avoidance of the various food deficiency diseases a simple matter. And modern communications can render all of this technical know-how readily available anywhere on earth. Yet half the world's population suffers from hunger and malnutrition.

The way out of our plight is clear, though the will so far has been grievously lacking. Population control is essential to survival. So too is genuine political and economic action by the rich countries to help the needy. Thirdly, we are going to have to curb our Western obsession with meat-eating — which diverts unreasonable hordes of food grain, fish and soya protein from human nutrition to animal fodder (and gives us far more protein than we really require). Fourthly, against an increasingly desperate world food scenario we must back all technologies, orthodox and heterodox, that promise to boost food production. Some ten per cent of the total land area on earth is presently under arable cultivation. Further areas are being claimed for agriculture each year, but the expansion of conventional crop and animal husbandry is clearly limited. About thirty per cent of the world's thirty-three billion acres is the most that can possibly be devoted to food production. The rest is too cold, too dry or too mountainous.

Evangelism on behalf of single, magic solutions to the world's food problem is rightly suspect. Unfortunately, brash salesmanship of that sort, backed by extravagant claims, has made many agriculturalists and nutritionists suspicious of one potentially invaluable stratagem — the mass cultivation of yeasts, algae and other microbes for direct consumption as food or animal fodder. Recently, too, promotion of such novel nourishment has been bracketed with a genuine absurdity — the marketing of such ersatz foods as soya beans processed to look like fillet steak. Ingenious but misguided applications of technology of this sort have nothing to offer in the context of the world's food predicament, and it is tragic that they should be proffered alongside microbial food. Steaks made of beans are an irrelevant absurdity. Microbial protein — produced by growing microbes on cheap and in many cases waste materials — could be a substantial complement to conventional food for both man and beast.

Typical of the way in which naive and insensitive technocrats parade their products as minor miracles is the sort of literature put out by some of the companies that are now beginning to invest heavily in microbial food production:

A major advantage of these processes is that the output of, for example, 15 fermentation protein plants covering a total area of not more than 20 hectares could replace 1.6 million hectares devoted to soya beans [declares one ICI pamphlet]. Western Europe, which takes about 45 per cent of world fishmeal supplies, could well have seven protein plants by 1985. The total fuel demand for these plants in one year will be about 5 per cent of that used annually by the population of Europe for private cars.

A comparison of microbial food factories with natural cultivation of soya beans is precisely the one that makes little sense, socially, politically or nutritionally — and indeed is likely to suggest that those who proselytize in this way simply do not understand the real world in which we live. It is enough to point out that we have grossly neglected the use of microbes as food, without suggesting that this new technology should in any way *replace* existing, well understood methods of raising vegetable protein. And to relate the costs of doing so to those of motoring comes close to comedy.

Nonsense of this sort should not, however, obscure the worthwhile contributions that microbial food can make towards relieving food shortage, particularly in the most needy geographical areas. Recent years have brought abundant evidence that global grain reserves are no longer adequate to balance the strains arising from clustered 'natural' shortages caused by droughts, floods and fishing failures such as the Peruvian anchovy disaster in 1972. We are going to be desperately short of nourishment in the decades ahead, and must explore all avenues that promise to ease the situation. As well as boosting microbial inputs into traditional agriculture, we should be developing with vigour the use of microbes as a direct, additional food source. ICI's fermentation plants will certainly not look pretty, and they are no substitute for traditional husbandry. But as gargantuan protein factories insensitive to the weather, such plants could play an important role in global food procurement.

We have also learned — agriculturally, economically, ecologically and energetically—not to put all our eggs in one basket. We have good reason to distrust singleminded propheteering from wherever it comes, whether from organic farmers, food technologists, green revolutionaries, or ersatz protein salesmen. Ecological diversity and the mixed economy are now seen as sensible principles over the entire breadth of human activity. Microbial food production, in that context, is no magic solution. Nor is it simply a glamorous product of technomania (as it appears when portrayed alongside hamburgers spun from reconstituted soya protein). It is a neglected but potentially handsome component of prudent food management.

The ample virtues of algae
Theoretically, the most exciting prospects for microbial food are offered by the autotrophs. They are self-sufficient

self-starters, creating sugars, proteins and other foods out of carbon dioxide gas and other simple inorganic materials (p. 34). Of all the autotrophic microbes, algae are outstandingly attractive. As well as high-quality protein, they produce fats, carbohydrates, vitamin C, B-group vitamins and other nutritional goodies — all from carbon dioxide. One of them, *Anabaena cylindrica*, is one of the richest known sources of vitamin B_{12}. The blue-green algae are valuable because of their high content of linoleic acid, the major polyunsaturated fatty acid, whose presence in the diet in place of saturated fats is thought to reduce the risk of athero-sclerotic heart disease. Algae secure the energy for their vigorous synthetic tasks by using solar rays — which they exploit more efficiently than do green plants. Some blue-green algae fix atmospheric nitrogen too, thereby assembling a rich assortment of foods from frugal resources. Algae can be cultivated continuously, not just seasonally or in batches, and may be grown in agriculturally useless, arid and waste lands. They do not even require excessive amounts of water. Finally, algae produce little or no uneatable waste.

Algae have been collected from freshwater and marine habitats in some parts of the world for many years. But the existence of such natural harvests has now been gravely threatened by the dissemination of DDT and other pollutants that we have jettisoned into the biosphere. There is also no doubt that artificial cultivation — where favourable conditions of nutrition, exposure to light and agitation can be maximized — will boost algal food yields prolifically. Work of this sort has been underway sporadically since the late 1940s and early 1950s, though only recently have world events conspired to stimulate research and development on the grand scale.

The single-celled alga *Chlorella* is one promising candidate as a micro-crop. When growing actively, its entire synthetic activity is mobilized to produce new cell material — as much as sixty per cent of which is protein, plus some thirty-five per cent carbohydrate and about five per cent fat. Some strains contain even more protein than an equivalent amount of dried beef or soya bean meal. Another species, *Scenedesmus,* is also potentially useful because it has a larger (30μ) and heavier cell than *Chlorella,* and can thus be harvested more easily. A third genus, *Tolypothrix,* which fixes nitrogen, has been cultivated (using hot spring water and natural gas as sources of heat and carbon dioxide) in Japan. It is grown primarily at the moment to inoculate into paddy fields, and thereby boost their fertilization, but it too has potential value as a food.

The earliest attempts to grow algae on a large scale seem very small-scale when judged against present-day thinking. The Arthur D. Little Company in Cambridge, Massachusetts, experimented in the early 1950s with cultures of 1 to 1000 litres. At about the same time scientists at the University of California tried out 10,000 litre cultures. Today, units of over 100,000 litres are operating in Japan, and the University of California has been experimenting with a million litre pilot plant. One recent calculation suggests that the entire protein needs of the US could be satisfied with 10,000 billion litres of algal culture.

At this point, heady unrealism takes over. Experts have weighed in with the reassuring notion that if we covered half the earth's land surface with tanks containing cultures of algae, the food so produced would support three thousand times the present world population. Others have pointed out that (should this prove insufficient) panels of algal culture could be floated on the oceans, thus making an additional seventy-three per cent of the earth's surface available to cultivate algae for human consumption. The fact that a world full of culture vessels would scarcely be worth living in escapes the attention of the algal prophets. And the need to deploy this type of bizarre wizardry is never seriously argued. What such calculations are good for, nevertheless, is to illustrate the staggering nutritional potential of the microbes. Though one would never wish to harness that potential on the astronomical scale envisaged by the wilder enthusiasts, their sums are helpful in highlighting the scale of our neglected resources.

Why, indeed, have we not yet exploited more vigorously such obvious potential? Curiously, the reason seems to be that, despite an abundance of free materials—carbon dioxide, sunlight and fecund microbes poised to generate trillions of microscopic, high-yield protein factories—we have been over-impressed by the likely financial outlay needed to establish routine plants to mass-produce algae. These costs cover such items as temperature control, agitation of the algae to ensure that all cells receive sufficient sunlight and the harvesting and drying of the crop. Now that we have become more conscious of the need to conserve resources, however, and to use virtually limitless commodities such as solar energy in preference to finite ones like fossil fuels, the picture has changed dramatically. Today algal culture is being assessed as an important and economically sound contributor to the planet's protein needs. Already algal culture is a reality in France, the US, Czechoslovakia, Mexico, Algeria, Japan and Uzbekistan. A process for growing *Scenedesmus*, developed at the Research Institute for Carbon Biology, in Dortmund, has been tried out successfully since 1971 in Thailand. The powder produced by harvesting and drying the alga looks like spinach but has a bland taste. It is highly digestible and a portion of vegetables containing 50 grams of *Scenedesmus* is equal in protein to a pork cutlet weighing 150 grams. A factory for producing the powder was established in India in 1973, and a second was completed in 1974.

There is especial interest in cultivating algae in sewage. More than a thousand kilograms of algae can be produced from 4.5 million litres of sewage, by growing the algae in shallow ponds. In a process now being developed in Thailand, growth of the algae is promoted by including bacteria, which liberate carbon dioxide and ammonia from the sewage. At the same time, the amount of offensive matter is greatly reduced. In other words, a method of sewage treatment has a dual function as a source of food. In this broader context, such a scheme is doubly compelling. The costs are genuinely low. And, being based on the idea of recycling, the whole arrangement is more consistent with long-term ecological stability than one that mindlessly consumes or jettisons resources without thought for the morrow.

On a small scale, the virtues of recycling involving algae are vividly highlighted by proposals to use them for life-support during long-distance space travel. The idea is that cultures of algae could both provide staple food for the astronauts and replenish the atmosphere inside the spaceship with oxygen by photosynthesis. Conversely, the human inhabitants would donate carbon dioxide by respiration, and possibly faecal material too. At least as far as Mars — beyond which there would probably be insufficient sunlight for the purpose — it seems entirely feasible to use algae for this purpose. Astronauts and algae would form a mutually supportive ecological community.

On earth, the chief future use for algal food will probably be as a contribution to protein in the human diet. Algae not only produce lots of this vital commodity, the protein is also of high quality — its content of amino acids is similar to that we require for making our own tissues. It contains, for example, a large amount of lysine, an amino acid that is in short supply in many cereal proteins. Moreover, though some otherwise desirable types of algae are deficient in particular amino acids, the algal farmer of the future will be able to choose strains with the most apposite pattern of amino acids for human consumption. Just as microbes used in the brewing industry have been improved, there is also great scope for geneticists to heighten the qualities of food algae, by exploiting mutation and other hereditary changes. As yet, this approach is almost totally underdeveloped.

People have, of course, been devouring algae since ancient times — in the form of the seaweeds — and this supports scientific evidence that the microscopic algae are not only nutritious but also perfectly safe to eat. Young stipes of *Laminaria* seaweed are consumed in Japan, and Pacific Islanders chop and eat several sorts of raw algae. Algal soups fed to patients in a leper hospital in Cabo Blanca, Venezuela, have been reported to be both palatable and nutritious. The US Army's Natick Laboratories are also conducting large-scale trials in which army personnel are consuming diets containing considerable quantities of algae.

As with other potential microbial foods, algae have sometimes been discounted as human food because they are enclosed in a tough and indigestible cell wall. Their critics have argued that the only conceivable use for them is as animal fodder — indeed, large-scale feeding experiments with cattle are now in progress at the University of California. Most animals accept large quantities of algae quite happily in their diet. Ruminants also digest the cells more completely than we can and thus use this novel food more efficiently. An alternative approach is to include fish in the ponds used to grow algae. The fish live on the algae, and though the final protein yield is lower than when harvesting the algae themselves, the process can be made highly efficient. And fish are more familiar than algae on the dinner table.

Clearly, there is potential here — though feeding algae to cattle or fish should not be seen as alternatives to human consumption. It is perfectly good sense also to use algae as a supplement in our own diets. We can make excellent use of algal protein, if necessary breaking the cell walls during production to release the contents. We consume many other plant materials encased in indigestible cellulose — which, moreover, constitutes roughage, a component of diet whose importance becomes increasingly apparent in maintaining the vigorous and healthy functioning of the gut and thereby preventing diseases such as diverticulosis. It is possible that the cell walls of some algal strains are a positive nuisance, being so indigestible as to cause intestinal upsets. If so, this can be overcome by using mutants that have fragile walls.

Much culinary contemplation and experimentation is now going into the task of devising interesting recipes for serving algal foods. The Japanese in particular have made considerable strides. Madam Tamiya at the University of Tokyo has developed recipes for using dried *Chlorella* in soups, yoghurt, ice-cream and bread. In Germany, at the Research Institute for Carbon Biology's algae factory in Dortmund, such tasty dishes as algal dumplings, algal noodles and mashed algal potato have been devised. All have been proven in practice. Civil servants at the Ministry for Economic Cooperation in Bonn have tried the foods and pronounced them not only edible but also appetizing. These are, however, small beginnings. Algae clearly have enormous, unrealized potential, both as bulk contributors to food requirements and as delicate novelties for the epicure.

Food from fungi

Like the seaweeds, macroscopic cousins of the microscopic algae, the large fungi have a venerable history as human food. The Romans and Greeks were familiar with the field mushroom and other edible fungi — which even today form a significant part of the diet in many parts of the world, including the Continent, Scandinavia, and Canada. Interest in the microscopic fungi as food dates from the development of the antibiotics industry in the 1940s and 1950s. Many of the microbes that produce antibiotics are fungi which grow as a network of filaments. After giving up their life-saving drugs, these filaments — often of high nutritive value — are left over as waste. There is little sense in simply destroying or discarding the mycelium. It can be fed to cattle, but only ingenuity is required to use it as human food.

Fungi produce protein at a phenomenal rate. The contrast with even those gross protein factories of conventional husbandry, beef cattle, is striking. An acre of corn yields some 1700 kilograms of grain, containing about 1500 kilograms of starch. If this starch is used as starting material for cultivating fungi, the eventual haul of protein, including that extracted from original grain, is about 270 kilograms. The equivalent yield, if cattle consume every bit of the corn (protein and starch included), is about 36 kilograms of steak protein. Calculations of this sort illustrate just how inefficiently farm animals convert their fodder into what we presently consider to be essential items of food. Compared with microbes, they also do so pitifully slowly. An average-sized bullock synthesizes less than half a kilo of protein every twenty-four hours. An equivalent weight of fungus, yeast or other micro-organism can manufacture some 50,000 kilograms of protein in the same time. Even a plant like soya bean does about a hundred times better than the bullock. Such figures give a new perspective to our thinking about food production, and must inevitably begin to influence both agricultural policies and thus eating habits in the future.

Times of war create the pressures, similar to those facing the entire world at the present time, which propel such fundamental points into the forefront of our thinking. It is no surprise that *Fusarium* and other fungi were incorporated into human diets during the last world war. Evidence gathered at that time suggests that people given these supplements enjoyed better health than those without fungus food. *Fusarium* is an even better source of protein than brewers' yeast. It forms the basis of a new microbial food being developed in Britain by Ranks Hovis McDougall, who grow the fungus on waste starch left over after bread manufacture and after the separation of protein from field beans. Another project that makes fungal food from waste material, in this case sulphite liquor from paper mills, is the Pekilo plant at Jamsankoski in Finland.

The ability to multiply on rubbish and vile effluents

is one of the outstanding attributes of microbes as food. Those we have so unjustly labelled *fungi imperfecti* are superb in this respect. They generate protein at an alarming rate from many different crude and cheap materials — sweet potatoes, wastes from the food-processing industries, cassava root — almost anything that contains some carbohydrate. Food production in this manner is especially relevant to developing countries, and in Britain Tate and Lyle are backing a 'village technology' project based on cultivating fungi on waste. One idea, for use in Central America, is to grow *Aspergillus niger* in a low-grade syrup obtained from citrus waste. The fungus is later harvested and used as animal feedstuff.

Fungi are also helpful in two-stage schemes that harness differing microbial skills in transmuting the indigestible into the digestible. The Swedish Sugar Corporation applies this principle in its 'Symba' process, which turns potato waste into animal fodder. The idea is simplicity itself. First, a fungus growing on the waste produces enzymes that convert the starch to glucose. Next, a yeast that cannot subsist on starch, but which manufactures a lot of protein and other valuable nutrients, is grown on the glucose. An especially attractive feature of the process is that the raw material is in ample supply and is an otherwise bothersome pollutant that would have to be rendered innocuous before being released into the environment. Looked at solely as a method of purifying effluent, the 'Symba' process costs at least ten times less per kilo of waste than conventional methods. The Swedish Sugar Corporation produces some one million kilograms of yeast per year by this method, and sells it as chicken feed.

Another intriguing application of fungi, as yet untried in practice but theoretically exciting, is to use their enzymic skills to transmute cellulose into starch. Starch and cellulose — both giant molecules — are composed of many linked units of glucose. But the two structures differ so greatly that separate enzymes are required to digest them — to break down the macro-molecules and liberate glucose. That is why the ubiquitous existence of vast quantities of cellulose — the basic building material of plants — is so infuriating in a world facing a food crisis. And this cornucopia of glucose is especially luxuriant in those very regions that suffer the cruellest famines and malnourishment.

To investigate means of bridging this gap, the US National Aeronautics and Space Administration and the American Society of Engineering Education set up a systems design project, whose findings were published in 1973. Conducted at Stanford University and the Ames Research Center, California, the project's aim was to come up with economically sensible methods of making starch from plant cellulose. Merely turning cellulose into glucose was not considered enough. Though digestible and nourishing, glucose cannot easily be incorporated in worthwhile quantities into staple foods. By far the most attractive proposal hatched by the analysts was one starting with sugar cane wastes (bagasse). After chopping and milling, the bagasse would be treated with an alkali to loosen the cellulose fibres, allowing impurities to be removed and permitting the ready access of enzymes produced by a fungus, *Trichoderma viride*. These enzymes would attack the cellulose, releasing glucose. Three further enzymes, from yeasts and bacteria, would then be brought into play to polymerize the glucose again — this time into starch.

A starch factory of this sort could be conveniently sited next to a sugar mill. It would produce a constant flow of starch for use in prepared foods, as an 'extender' in conventional foods, or as animal fodder. The design team point out that starch is already widely included with flour in baked goods in many parts of the world, and recommend this as the most appealing use for starch from such a plant. Many foods containing little else but starch are also common, such as 'Instant Ramen', a noodle packed in individual servings, which is popular in Japan. The potential clearly exists, but the scheme devised at the Stanford-Ames project remains practically unexplored.

Fungal assistance in turning cellulose into food is peculiarly welcome in developing countries, because fungi can be found that will release glucose from cellulose in whatever form it comes. Speaking at the Australian Institute of Food Science and Technology in June 1974, Professor T. K. Ghose of the Indian Institute of Technology, Delhi, pointed out that, as well as innumerable plant materials, old newspapers would serve as food for *Trichoderma viride* — which should be able to manufacture glucose from this unusual source at a price comparable with that of the natural product. Enzymes from the fungus can turn newspaper into a solution resembling 'thin honey' in twenty-four hours. At that stage, Professor Ghose favours using the glucose to grow yeasts for human consumption, rather than the more sophisticated idea of turning it, via microbial enzymes, into starch. Research workers at Cardiff University in Britain, who have developed a similar protein-from-paper process, believe that the protein equivalent of a steak weighing over two hundred grams can be made from three hundred grams of paper.

Yeast eating, today and tomorrow

As well as being the first to be consciously recognized, yeast is one of the most promising forms of microbial food currently being evaluated. Though brewing is many centuries old, only during the late nineteenth century did we realize the value of the left-over waste yeast. Cautiously, it was added to livestock feeds. Then, in 1910, Max Delbrück and his colleagues at the Institut für Gärungsgewerbe in Berlin investigated the suitability of yeast for human consumption and developed industrial methods for propagating it. Their work was further stimulated by discoveries of yeast's nutritive value —

its richness in vitamins and the capacity of yeast protein to replace vegetable and animal protein. Casimir Funk, who discovered niacin (one of the B group vitamins) in yeast, was the man who originated the term vitamin. As a prolific reservoir of B group vitamins, yeast was crucial in helping to eradicate deficiency diseases such as beri-beri and pellagra between the wars.

Economic circumstances before and during the Second World War brought increasing interest in food yeast, particularly in Germany and Britain. Only afterwards did allied inspection teams uncover the processes developed at the Institute für Gärungsgewerbe for growing *Candida utilis* yeast in the sulphite waste from the paper pulp industry. At least eight food yeast plants were operating in Germany at the height of the war, producing some sixteen million kilograms of *Candida* that was incorporated in human food. Most used pulp-mill waste, though one of them cultivated yeast in dairy effluent. Successful efforts were also made to raise yeast as a source of fat by growing specially selected strains. British scientists opted for molasses as the starting material, and in 1944 the UK Colonial Office opened a commercial yeast plant in Jamaica. The project was short-lived, however, mainly because the food yeast was somewhat unpalatable owing to traces of a defoaming agent added during manufacture. Also, efforts to explain the merits of the yeast to potential consumers were woefully inadequate. Production of food yeast from sulphite liquor at Rhinelander, Wisconsin, which was initiated during the war, also did not survive as a continuing scheme.

Today, triggered by food and resource crises, companies and research institutes throughout the world are again vigorously exploring food yeast manufacture. World annual production of food and feed yeast is already around 250 million kilograms. Much of this is in eastern Europe, particularly in Russia, where output is expanding very rapidly. Taiwan, South Africa and Puerto Rico also have major facilities producing food yeast. Molasses and sulphite liquor have been the principal raw materials to date, though agricultural wastes, fruit juices and even dried leaves and faeces have been used. The economic merit of such processes is clear. A young chicken or pig takes at least a month to double its weight. A yeast cell does so in less than two hours. A moderately sized yeast factory can thus maintain a continuous output of 10,000 kilograms of yeast per day. To make the equivalent amount of protein would mean killing eighty pigs each day.

Much interest now centres on using petroleum and natural gas as starting materials. BP was the first company to investigate seriously the possibility of raising yeast on hydrocarbons. Dr Alfred Champagnat was their man who, in the late 1950s, began turning a longstanding piece of academic knowledge — that yeasts can thrive on various fractions of petroleum — into the technology necessary to apply this information on the grand scale. As a result of his vision and leadership, BP now has two pilot plants in operation. One is at the Société Française des Pétrole's refinery at Lavera near Marseilles. The other is at Grangemouth in Scotland. A third and much larger plant, a joint venture with the Italian company ANIC, opens at Sarroch in Sardinia at the end of 1975.

Gas oil is the starting material for the Lavera factory. Gas oil is a heavy petroleum fraction, not unlike domestic fuel oil, which contains some ten per cent of waxy paraffins. The yeast *Candida lipolytica* uses these paraffins to grow — and at the same time helps to refine the petroleum, the remainder of which is returned to the conventional refining process. The final yeast product contains over sixty per cent of protein — about the same as dried fish or meat, and double that of dried skim milk. A major plant to manufacture yeast by the same process is also under construction in Czechoslovakia.

BP's Grangemouth unit uses a slightly different method, which will also be the one adopted at Sarroch. Here the yeast grows on paraffins that have already been purified from raw petroleum. The advantage is that the microbial cells can be harvested afterwards without further processing. Yeast made at Lavera has to be treated to remove unwanted gas oil — and this also extracts some of the fats in the cells. The net cost of the two processes is about the same, however, and BP considers them to be complementary.

What to do with the vast amount of yeast generated by such plants is as much an enticement to culinary skills as is the production technology a challenge to the industrial microbiologist and engineer. At present, while some people consume yeast in compressed forms as a health food, most of us also eat it (usually unknowingly) as a constituent of many everyday foods. Smoked yeast, which gives a bacon-like flavour and taste, is especially popular with food manufacturers. Powdered yeasts are added not only to rissoles, sausages, breads, buns and rolls, but also to macaroni and noodle products, as enriching sources of vitamins and minerals. Biscuits, doughnuts, breakfast cereals, baby foods, peanut butter, muffins, meat dishes and many snacks also frequently include a small proportion of yeast. 'Incaparina', the popular cereal mixture eaten by people in Central America, contains three per cent of yeast. Introducing any novel food (microbial or otherwise) is always a tricky business, and many crusading technocrats have learned to their cost how foolish it is to ignore social and psychological factors influencing food choice. But with yeast the producers have a head-start. Many folk already like to munch yeast; most of the rest of us, without realizing the fact, appreciate yeast flavours in various foods. Deriving a higher proportion of our daily protein from yeast could be not only acceptable but also enjoyable.

One question has loomed recently over the yeast-from-petroleum lobby — whether such processes can remain attractive despite unprecedented rises in oil prices. It seems, in fact, that making food yeast in this manner will remain a competitive option. Firstly, the amount of oil needed is miniscule. Professor Tokuya Harada has calculated that an annual production of 7,500 million kilos of yeast protein — enough to overcome the global protein shortage predicted for the year 2000 — could be achieved by using only five per cent of the world's total present annual petroleum consumption. Secondly, the manufacturing processes will be made more efficient; the technology is new, and considerable improvements are inevitable in the light of experience.

Thirdly, and most importantly, oil is not the only commodity that has risen sharply in price over recent years. The price of sugar, the logical alternative, for growing food yeast, has climbed steeply since the termination of the International Sugar Agreement in 1973 and the ensuing world shortages. And animal feedstuffs have become much more costly. That is the crucially important comparison. Seen in broad perspective of financial and energetic cost, beef husbandry and microbial food production (even using oil, whose price has rocketed from its former unrealistic low) lie at opposite extremes of the spectrum. Thanks in each case to the microbes, both will doubtless continue in the decades ahead. Beef, however, will rise inexorably in price, just as surely as yeast and other microbial foods will become an increasing part of day-to-day cuisine. The net result will not be to impoverish our enjoyment of food, but rather to extend and promote it through greater variety.

BP scientists are not the only believers in the future of protein production from hydrocarbons. The Russians have opened a plant making yeast protein from purified liquid paraffins. Gulf Oil is working with a similar large-scale process at Wasco, California. Liquichimica SpA is building a plant at Saline di Montebello in the Calabria area of Italy, using know-how supplied by Japanese technologists. Japan was one of the earliest countries to announce plans for growing yeast on oil fractions, but encountered early setbacks when controversy arose about the potential danger of contaminants in yeast produced in this novel fashion. Towards the end of

1974 Mitsubishi Petrochemical announced a process for manufacturing *Candida* by growing it on (surprisingly) ethyl alcohol — of which Mitsubishi is the major Japanese producer. It is likely, however, that this and other Japanese companies will soon be making food yeast in a big way from petroleum feedstock.

Safety is of paramount importance before any novel food can be made publicly available. Enormous numbers of toxicological tests are now in progress to rule out possible hazards of the emerging microbial products. While all such foods have to be cleared, a greater degree of suspicion must rest initially with microbes raised on unusual substrates, such as petroleum, compared with those reared on 'traditional' materials like molasses. Moreover, while BP's yeast is a natural strain, there is enormous scope for microbial geneticists to evolve improved varieties, with higher protein content, for example, or more rapid growth. The vetting of such strains will clearly have to be even more stringent than with more familiar microbes that have probably co-existed with us for centuries. It was in part to lobby for consistent regulations on safety that the European Association of Single-cell Protein Producers came into being at a meeting in Brussels on 27 November 1974.

Bacteria from oil and gas

Interest in bacteria as potential food sources dates from the early 1950s, when tests showed that a sludge made by growing the intestinal bacterium *E. coli* in mass quantities was an excellent protein supplement for chicken and rats. Bacteria have two appealing characteristics as food. The first is their unsurpassed growth rate (some forty times greater than algae and about four times greater than yeast), permitting the rapid manufacture of vast quantities of cell material. Second is the quantity (up to eighty per cent or more) and quality of their protein. Bacteria tend to carry appreciable amounts of the sulphur-containing amino acids, which are sometimes deficient in both fungi and yeasts. The lesson here may be to work towards a microbial cocktail, in which the strengths and limitations of the amino acid content of their respective proteins complement each other. Symbiotic mixed cultures of yeasts and bacteria might yield highly nutritious food. Tests have already indicated as much for yeasts and lactic acid bacteria grown in whey, the waste material produced in huge amounts by cream cheeseries.

Bacteria also commend themselves because of their skills in living off virtually any material with which they are presented that contains those essential elements, carbon and nitrogen. Species of *Streptomyces*, for example, can digest such unpromising waste matter as hooves, wool and feathers. Vast quantities of this rubbish, rich in nitrogen, are generated and discarded every day by industry. It could be exploited as a source of food. Another bacterial specialist, *Hydrogenomonas*, lives on three gases; carbon dioxide, oxygen and hydrogen. It is being evaluated as a potential food source in areas well endowed with hydroelectric power, where hydrogen and oxygen can be generated cheaply by passing electricity through water.

The two starting materials being investigated most intensively, however, are methanol, which is mass-produced from oil, and methane, the major constituent of natural gas. In Britain ICI began looking seriously at these substances in 1968. After initial studies with methane, the company has since developed an efficient process for cultivating edible bacteria on methanol. Though methane is cheaper, ICI considered that the explosion risk and the difficulties of growing bacteria on a gas rather than liquid were such as to swing the choice in favour of methanol. Having selected the feedstock, ICI scientists tried out hundreds of different microbes to find one that thrived well on methanol and was hyper-efficient in converting it to protein. The bacterium they eventually chose, a pseudomonad, has an interesting background. A team of microbiologists at ICI's Corporate Research Laboratory at Runcorn, Cheshire, had been assessing the prospects for employing bacteria as sophisticated chemists, to turn cheap bulk chemicals into rare compounds whose synthesis would otherwise be costly and time-consuming. One strain, which they discovered living in the soil underneath an industrial methanol plant, proved to be an avid methanol consumer. It was this bacterium that ICI's protein group later selected for the star role in its new food process.

An initial snag in cultivating masses of bacteria is that they are more difficult to harvest than are yeasts, chiefly because of their smaller size. ICI has solved this problem. The pseudomonad grows inside a specially designed vessel, with no moving parts, and injected air stirs and vigorously circulates the methanol and proliferating bacteria. A 'soup' containing the microbes passes from the vessel and is treated by a secret process which aggregates the cells, before they are fed as a slurry to a centrifuge for further concentration. The final product is of high nutritional quality. ICI has conducted trials to compare the microbial protein with conventional foods such as soya and fishmeal for pigs, poultry and calves, the bacterial protein being increased gradually until it comprises ten per cent of the total. There was no significant difference in the animals raised on the novel and conventional diets — except for one set of experiments in which pigs given the microbial food showed a significant improvement during the final growing period compared with those on the ordinary diet. In 1973 ICI commissioned a £1 million plant, with a capacity of some one million kilograms of protein per year. Largely because of the social and psychological difficulties involved in introducing new foods for human consumption, however, the company is at present interested only in the animal feedstuffs market as the outlet for its novel protein.

Another company, Shell, has persevered with methane as starting material for making microbes. Methane is certainly an enticing substrate. Not only does it occur in

natural gas — North Sea gas is almost pure methane —
but also, if such reserves were exhausted, it could still be
made inexpensively by chemical methods. Shell's interest
dates from the sudden discoveries of the previously hidden
riches beneath the North Sea in the mid-1960s. By
October 1974, after spending a million pounds on research
over the previous nine years, the company was able to
announce that it had successfully developed a process
and operated a pilot plant. There is no single star microbe
here, but rather a mixed cast containing at least two different strains. The organisms grow in a solution, containing ammonia and other nutrients, through which passes
a mixture of methane and air. A thin suspension of cells
is drawn off continuously and passes to a separation process which yields a soft cheese-like mass. After drying,
this becomes a creamy-white powder with very high
protein content (seventy-five per cent). The amino acid
spectrum of the protein is excellent too, approaching that
of best-quality white fishmeal. By the early 1980s Shell
hopes the new product will be on the market and will be
competitive with fishmeal and soya bean feedstuffs.

Costs notwithstanding, natural gas and oil are destined
to play a key role in future production of microbial food.
They have an additional advantage over waste vegetable
materials as media for cultivating microbes — and certainly over traditional fish and vegetable sources of food.
They are not subject to seasonal variations in supply.
Moreover, even hydrocarbons are wasted at present. Some
100 billion kilograms of natural gas associated with oil
is simply burned — 'flared-off' — every year throughout
the world. This could be used to support a Shell-style
microbial food factory. Alternatively, the methane might
be converted to methanol and then fed into an ICI-type
process. Even in times of global energy crisis, it is simply
not economical to liquefy the methane and transport it
to areas of need. Yet ICI calculates that it would cost about
the same to construct a methanol plant at the methane
source as it would to liquefy the gas. The resulting food

would then be sufficiently valuable to bear its transpor-
tation costs.

Assured future for microbial foods

Until recently, the story of microbial food was simply one
of academic interest before the last war, followed by frantic
efforts to apply that knowledge under the exigencies of
war, followed in turn by apathy and neglect. Just as we
were profligate in our use of energy in the 1950s and 1960s,
before the shocks of the early 1970s, so we felt able to ignore
the enormous potential of the microbial world in boosting food supply. Today, great resources of money and
manpower are once again being thrown into research
and development in this area. Eating microbes will not
solve the food crisis. We shall have to harness micro-
organisms more effectively in many other subtle ways
in conventional agriculture if we are to come anywhere
near to repelling the spectre of famine. But microbial
food is one vital item on the agenda for the next few decades. It stands out all the more vividly because of its
abundant promise and its past neglect.

In forecasting future developments in technology, the
'Delphi technique' has proved both successful and stimulating. It is based on carefully weighted guesses and
judgements from a panel of experts given a questionnaire
covering trends in the area under examination. The results
of such a study carried out anonymously at the University
of Reading and published in 1972 cast interesting light
on the future of microbial foods. They showed that, while
novel proteins generally were then insignificant in relation to the total protein consumed in the United Kingdom,
rapid growth in both production and acceptance was
expected over the decade. Though the panel highlighted
an existing protein, extracted from soya bean, at the top
of its list of 'new proteins' likely to be consumed in 1981,
it ranked fungal and yeast foods next, above leaf protein,
fish protein concentrate and the six other versions considered. In all, the novel proteins were expected to have

, captured ten per cent of the meat and meat products
market and ten per cent of the animal feedstuffs market
by 1981. A decade later, the panel predicted, the corresponding figures would be twenty-five per cent.

Similar confidence is apparent from a five-volume
report on the future of microbial protein in animal feedstuffs, published late in 1974 by the Wolfson Laboratory
for the Biology of Industry, Cardiff. It estimated that by
1980 European calf-, pig- and poultry-breeders could be
buying no less than £50 million worth of microbial protein
to supplement their fodder. The report argued that,
though microbial food production is not yet a proven,
established technology, the remaining work is largely
that of engineering rather than development. As well as
confirming the economic feasibility of turning waste gases
in the Middle East into food, the Wolfson study suggests
that sewage and even newsprint could become widely
exploited sources of microbial food.

Textbooks need re-writing
Most textbooks of microbiology contain a chapter entitled 'food microbiology'. Most of those chapters are
devoted entirely to the few microbes that cause spoilage
of food (usefully so, by the way, because such spoilage —
usually a result of human error — indicates the possibility
that a pathogenic organism may have entered the food
too). With the exception of a passing reference to bread
and sauerkraut, the positive uses of microbes themselves
as food are ignored. The bizarre contrast between this
omission and currently burgeoning developments in
microbial food production is a measure of the pace of
technological and political change in the past few years.
The textbooks are going to have to be rewritten.

7 Microbes as Scavengers

I once worked for a chemical company that had a reputation as a local nuisance. The problem was a nasty smell, escaping from a plant making amines — chemicals that are among the most malodorous known to man. The company was trying to put matters right, but with little success. One challenge was what to do with the liquid effluent. This was no more than a dilute solution of amines — so dilute, the company argued, as to make it uneconomic to try to recover the chemicals from such large volumes of water. On the other hand, the effluent could not simply be discharged into the nearby river (at least not knowingly) because of the horrendous stink it created.

Chemical methods of purification had been tested. My job was to help in trying to devise a cheaper, microbiological technique, and my first experiment began by filling a tall glass cylinder loosely with glass beads. Then I arranged for a solution of one of the offensive amines to percolate down through the beads, before being pumped back to the top of the column. The amine thus flowed slowly in a continuous, closed circuit. Next I took a fistful of soil, and occasionally added a pinch of it to the beads at the top of the column. The set-up was then left to tick over for several weeks. From time to time I analysed the liquid dripping from the bottom of the column to see how much amine it contained. There was no change at first. Then, after a week or so, the amine concentration in the fluid began to fall steadily. Eventually, the malodorous material all but disappeared from the water. The liquid percolating through a similar apparatus, to which I had not added soil, retained its original amount of amine. It sounds simple, and so it was. Much more work was necessary to develop an effective routine process for treating the effluent, but the original observation was child's play.

Microbes were responsible for what happened. From the varied and teeming microbial life in the soil sample, bacteria had emerged which were able to break down the noxious amine. Growing on this bizarre food, the bacteria had begun to proliferate and to establish themselves on the glass beads in the cylinder. When I renewed the amine in the liquid, that too was efficiently eradicated by the scavengers in the column. Had the unwanted substance been another amine, or some totally different waste chemical, another microbe in the soil would probably have been equal to the task of degrading it. Such is the metabolic versatility of microbes — particularly of the bacteria and fungi that live in the soil. Microbial systems of waste disposal have correspondingly catholic criteria of acceptance. Virtually the only proviso is that the material to be rendered inoffensive should be present initially as a relatively weak solution — not in such strength that it kills the micro-organisms before they are able to start work.

What is remarkable about the scavenging abilities of microbes is that the environment contains an apparently limitless reservoir of specialists capable of breaking down not only plant and animal materials but also an ever-increasing profusion of man-made chemicals. After billions of years of activity in building up and sustaining the great cycles of nature, bacteria and other micro-organisms are now helping to cleanse the environment of products of chemical industry that have never previously existed in nature. So far, science and technology have spawned about a million different chemicals and preparations that are used in industry and the household, and we add over 100,000 products to this list each year (often without good reason). Many if not most of them find their way into the environment. As Dr Stanley Dagley puts it: 'A visitor from another and, hopefully, less polluted planet might well be astonished that we blithely scatter novel compounds over the face of the earth, while leaving in comparative neglect a study of the capabilities and limitations of the microbes necessary for their removal.'

The fact that microbial ingenuity has been adequate to cope so far with our mindless jettisoning of chemicals in no way justifies us in ignoring and merely trusting in these unseen talents. What we ought to be doing is seeking to understand much more fully the machinery command-ed by microbial scavengers, and to harness their cleans-ing skills consciously and intelligently. Some steps have been taken in this direction recently, but the assured reservoir of ability, coupled with our present ignorance, suggests that considerably more could be done. The microbes are not only willing slaves: they are uniquely inexpensive ones, able to subsist and indeed thrive on materials that we find dangerous, inconvenient or repugnant.

Putrid into pure

Just as yeasts can either churn out great volumes of a beer-type drink by continuous fermentation or create the most inestimable clarets, so the scavenger microbes range from industrious mass disposal agents to sophisticated spe-cialists with the enzymic skills to dissect a single type of molecule. The outstanding example of large-scale scav-enging is sewage disposal — a process whose principles *are* understood and which has been a central feature of post-nineteenth century hygiene and sanitation. Its importance in modern society is reflected by the panic headlines that appear in the press whenever sewage dis-posal is interrupted — as in 1974-5, when a strike in Glas-gow led to vast quantities of raw sewage being dumped into the Clyde, and the ensuing gross fouling and health hazards caused a bitter local furore. If industrial bargain-ing strength is directly proportional to the vulnerability of society to strike action in its various sectors, then sewage workers rate very highly indeed.

But the real workers in sewage works are microbes. Tended occasionally, they are prepared to labour continually, turning the foulness of sewage into sparkling, pristine water. The complex network of processes that effects this transformation constitutes one of the genuine wonders of the world. What sewage microbes are presented with is a vile mixture of human and animal faeces, urine, vomit, menstrual flow, hair and other fibres, condoms, toilet paper and sanitary towels, oily filth off the roads, dirty soapy water from laundries, unmentionables from hospitals and abattoirs, liquids and solids discharged by the painter, the farmer and the garage-man and an unpleasant profusion of other grot washed into the sewers by the rain and through industrial and domestic sinks, sluice-rooms, baths, lavatories and outflows. All of this is in the early stages of foetid decay. The microbes accept the challenge and, unfailingly and efficiently, turn the devilish cocktail into water pure enough to discharge into the cleanest of rivers — or to treat and chlorinate for use as drinking water. Virtually the full range of micro-organisms — bacteria, fungi, protozoa and others — is required to complete the work. Innumerable different species contribute to the overall goal. But the mammoth and unceasing task is nonetheless accomplished with quiet competence. Even in these days when we are having to re-use water more frequently, sewage microbes are entirely capable of rejuvenating it in a form that we find acceptable once more. The scale, complexity and efficiency of the process is difficult to visualize by anyone who has ever visited a sewage farm or glanced at one from a passing train — when nothing ever seems to be happening.

The principal stages of sewage treatment approximate to those of the great organic cycles in nature. Complex proteins and other compounds, both soluble and insoluble, are broken down into simpler substances. Nitrogen incorporated in organic molecules is converted to ammonia, and their carbon is released as carbon dioxide. As in the soil, the ammonia is turned into nitrate. In addition to this natural sequence of transmutations, countless special microbial manoeuvres degrade man-made substances, including detergents, which nowadays find their way into sewage. The organic matter destroyed during sewage disposal consists of both dead material and also microbes such as *E. coli*, which enters sewage in astronomical numbers from the human intestine. The sewage scavengers deal just as efficiently with the typhoid fever bacillus and other occasional pathogens. It is the labours of the scavengers that make it impossible for typhoid and cholera to spread as they used to when we dumped our sewage in the streets and drew our water from contaminated wells.

If untreated sewage is poured into a lake or stream, not only does this pose a potential public health risk because of the pathogenic microbes it may contain: the sewage also creates a dreadful smell. This comes from both its overtly nasty ingredients, and the action of microbes that devour oxygen, generating anaerobic conditions that encourage the imbalanced proliferation of putrefying organisms which yield hydrogen sulphide and similar malodorous products. The water is ruined for drinking and recreational purposes, and will not support the growth of fish or other forms of aquatic life.

Sewage disposal usually begins by exploiting the very same processes that can precipitate these unwelcome consequences when we treat the environment with such foolhardy contempt. But at the sewage works, they are harnessed in a controlled way — one that causes no unpleasantness even for the operators. The gist of the procedure (after initial screening to remove bottles, pieces of wood and other solid debris that should not be there anyway) is a complicated series of microbial reactions. Some resemble those of ruminant digestion. Others are closely akin to alcoholic fermentation. Different organisms take part in this operation, which from such complex starting materials as fibres and cellulose yields two gases, carbon dioxide and methane. At least four sorts of scavengers play distinct roles in the overall process. Some digest the insoluble matter, using their enzymes to release

soluble substances. Other microbes ferment these products to alcohols and acids, which in turn are fermented
by a third group of microbes to carbon dioxide and hydrogen. Finally, specialized strains of bacteria combine some
of the hydrogen and carbon dioxide, yielding methane.
The entire routine goes on semicontinuously in large
enclosed tanks. Fresh sewage is introduced from time to
time, and the end-products drawn off. What emerges is a
solid residue and a liquid effluent, greatly diminished
in organic content by the conversion of so much of its
original burden into gaseous form. The solid detritus
contains indigestible matter and bacterial cells.

Though not particularly rapid, the process is stagger-
ingly efficient. A piece of linen fabric, attacked by bacteria, can disappear totally in five to seven days during
sewage digestion. The active microbes include known
species that demolish all manner of organic matter, from
paraffin to rubber — including even disinfectants such as
phenol. Many others are not well understood. Despite
our heavy reliance on its effective operation, the *dramatis
personae* of sewage disposal is still far from completely
delineated. We do not even know whether a typical sewage
works is 100 per cent or only five per cent efficient.

Little if anything need be wasted during sewage disposal — which we are beginning to recognize not simply
as a necessary chore in coping with obnoxious effluvia,
but also as a source of riches. Instead of simply burning
the sludge from digestion, many sewage plants now dry
it and sell it as fertilizer or lawn dressing rich in nitrogen,
phosphorus and potassium. Milorganite, widely used in
the US, is one such product. Another is Morganic, made
at the huge Mogden works at Isleworth near London.
Sludge is also finding novel applications such as the re-
clamation of land ruined by strip-mining.

The other invaluable resource is methane, which comprises about seventy-five per cent of the gases evolved
during digestion, and can be piped off and burned. Britain
leads the world in harnessing this as a source of energy.

About two-thirds of the country's five thousand sewage plants generate methane that is used for heating. Methane from the Mogden works is incorporated into London's gas supply, and some is sold to the British Oxygen Company. In the particularly efficient 'heated digestion' system, the methane often not only powers the engines operating the pumps but also warms up the digesting mass to a high temperature at which the microbes operate at peak efficiency. Yet many sewage works merely burn off their spare methane, without harnessing the heat so generated. No one has investigated the possibility of converting the methane into protein, by the sort of process we considered in the last chapter. And throughout the world large quantities of untreated sewage are still being jettisoned into rivers and the sea. In Britain over a thousand miles of rivers are grossly and dangerously polluted with untreated sewage.

The prospect of turning human faeces into useful and valuable products has also become popular with environmentalists and exponents of 'alternative technology'. They see small-scale decentralized technology as desirable in itself, and thus advocate domestic rather than centralized sewage disposal. Various forms of lavatory have been devised that exploit anaerobic digestion to produce compost, which is used as fertilizer, and in some cases methane that can be burned for heating. 'Composting loos' require little water, and they are efficient in conserving the nutrients in sewage and returning them to the soil. Another idea that has also been proved in practice is the generation of methane from farm manure by microbial digestion. The process takes place in oil drums and the gas is stored in bottles. Methane replaces petrol as a fuel for the internal combustion engine and can thus be used to power cars and farm vehicles.

Sewage farms also exploit microbes that thrive in the presence of air. The familiar trickling filter is simply a bed of crushed stones or coke, about two metres thick, on top of which the fluid containing disposable material is sprayed, often through rotating arms. The liquid may be effluent from a digestion process, or it may be raw sewage. A mixed population of microbes establishes itself on the rocks, and as the sewage trickles downwards air perfuses upwards through the filter. Filamentous fungi and bacteria, together with slime-forming bacteria, remove organic matter from the sewage and also help to bind together the microbial film covering the rocks. Algae assist by supplying oxygen. Gradually, some of the microbes are eaten by protozoa, which in turn are consumed by larger creatures. The net result of this food chain is the removal of organic material in the trickling sewage and its conversion, through respiration, to carbon dioxide.

Meanwhile, other chemical conversions are occurring. As specialist microbes pull apart and oxidize the organic matter, nitrogen is released as ammonia, which other bacteria commonly active in soil transform into nitrate. Similarly, organic sulphur first becomes available in the form of hydrogen sulphide, which autotrophic bacteria change to the much less offensive sulphates. Thirdly, further organisms extract the phosphorus from nucleic acids and turn it into phosphates. These three types of salts — nitrates, sulphates and phosphates — are still considered to be nuisances in the effluent from sewage plants, because they are such excellent nutrients. One snag is that algae thrive on them, causing the 'problem' of algal blooms in rivers receiving the purified effluent. Unduly high levels of nitrates in drinking water also create possible health hazards. The solution, of course, is not simply to throw nitrates away, as at present, but to make use of them. Much research is proceeding on methods of doing so — including the possibility of consciously cultivating algae in sewage fractions as a source of food or fodder (Chapter 6).

A different but widely adopted technique for disposing of sewage aerobically is the activated sludge system. Large volumes of compressed air or oxygen are forced through sewage in a tank. Suspended particles flocculate after a time into tiny gelatinous masses swarming with microbes that thrive on the organic matter, breaking it down speedily and efficiently. The floccules are called activated sludge. The key role in their activity is played by *Zoogloea ramigera*, a bacterium that forms slime to which protozoa and other microbes become attached. The amount of sludge gradually increases as air and sewage circulate through the tank, causing a sequence of chemical changes similar to that in a trickling filter. By adhesion and adsorption, the sludge gathers to itself much of the microbial life, colloidal material, colour and smell of the original sewage. Eventually, the fluid passes to a settling tank, and some of the activated sludge that precipitates is returned to the tank to prime the process once more. The rest is removed, and it too may be dried and sold as fertilizer. Like the product of a trickling filter, effluent from an activated sludge system is rich in nutrients and thus favours the growth of algae. A modified technique, announced in 1974 by Britain's Water Pollution Research Laboratory, Stevenage, uses an extra anaerobic stage, during which bacteria convert nitrate to nitrogen gas, thus lowering the nitrate content of the effluent by eighty per cent — and at reduced running cost.

Finally, some sewage disposal is based on lagoons — large, shallow, open lakes that can be used to deal with either raw sewage or partially treated wastes. They are particularly suitable in areas where land is cheap, and are especially efficient in removing viruses from sewage. Oxygen either diffuses from the air or can be pumped in mechanically. Algae and other tiny aquatic plants tend to grow prolifically on the surface because the lagoons are open to the light. Depending on the shape of the lagoon, the processes may be totally aerobic or partially anaerobic in the depths. Bacteria proliferate, consuming oxygen and producing carbon dioxide while destroying and breaking down the organic matter. The algae photosynthesize, fixing carbon dioxide and generating oxygen. Bacteria and other microbes attack moribund algal cells, converting their vital molecules to carbon dioxide. As a balanced

ecological system, it is ruthlessly effective in destroying
the organic material in sewage — especially when relatively small quantities have to be treated. Unlike most similar
methods, it does not usually generate large amounts of
soluble nutrients. Towards the bottom of deeper lagoons,
bacteria promote denitrification — the reverse of the
normal transformation of ammonia to nitrate. This releases nitrogen and ammonia as gases, so diminishing the
nitrate concentration in the water.

Sewage lagoons are often harvested, giving as much as
one to five thousand kilos of algae per acre per month.
That is a considerably bigger yield than with most farm
crops. The algae can be used as fodder or digested anaerobically to generate methane. Professor William Oswald,
at the University of California, who is working on techniques of this sort as an efficient means of tapping solar
energy, has calculated that the energy content of algae
grown in domestic sewage is equivalent to about six
kilowatt hours per kilo. This was based on the lush algal
growth achievable in California on domestic sewage (or,
better still, chicken manure). But, given sufficient land to
construct large lagoons, the process can generate useful
amounts of methane in many less favourable climes. A
study conducted by the European Economic Commission
in 1974 showed that the solid wastes from a city of two
million inhabitants could supply sufficient material to
feed a 1000 megawatt power station, after microbial conversion. That would provide electrical power for some
80,000 people.

Whatever the process, an outstanding and amazing
characteristic of sewage organisms is their capacity to deal
with whatever chemical comes along, whether as a regular
component of sewage or as an occasional passenger. The
exceptions — the most famous of which in recent years was
the foaming caused in rivers by detergents that passed
unchanged through sewage works — are conspicuous
and well known simply because, for the most part, sewage
microbes cope competently with all comers. In the case

of detergents, which even interfered with the smooth running of the disposal system itself, foaming troubles led to legislation requiring manufacturers to market only detergents that could be broken down by the micro-organisms in treatment plants. This they achieved by changing slightly the shape of the detergent molecules. The same principle could well be applied to other materials.

Explaining scavenging versatility

Few microbiologists have followed Serge Winogradsky's injunction to study the mixed populations of microbes as they occur in nature. This is one reason why the details, in contrast to the broad outlines, of sewage disposal have been so cursorily investigated. (The other is microbiologists' reluctance to scrutinize a process which, discovered and applied largely empirically, has worked superbly without scrutiny.) So we understand little about the micro-machinery responsible for the emergence of novel scavenging skills, whether in a trickling filter or a column carrying glass beads and noxious amines. We do know that a new compound is often broken down only very slowly at first, but that after an interval decomposition proceeds more rapidly. Added anew some time later, the substance is then degraded briskly from the outset, without a period of adjustment.

This suggests that time is required initially for the emergence, out of the varied micro-population, of an organism with suitable talents. Given both the need and the opportunity to devour an enticing new pollutant, that microbe would then proliferate more actively than its colleagues, up to the point where a sufficient number of cells were available to complete the degradation. Alternatively, a mutant may emerge at just the right time — an organism uniquely capable, like a microbial Churchill, of tackling the task in hand. But in some situations there may be more to the story than that. Dr J. S. Waid, a botanist at the University of Canterbury, New Zealand, believes so.

He is particularly interested in herbicides such as 2, 4-D which (amazingly, as complex substances that almost certainly never existed before being synthesized by the modern chemical industry) are vigorously decomposed by soil and sewage microbes. After a small initial application of 2, 4-D to the soil, a microflora develops that can very quickly degrade a much larger concentration later on. It is difficult to explain an effect of such speed and magnitude on the basis of selection, mutation or the adaptation of chemical machinery already existing inside the microbial cells.

Dr Waid has suggested that the spread of plasmids (transferable genetic elements, p. 30) may be responsible. Laboratory studies have shown that the frequency with which such plasmids are transferred is higher when bacteria meet on solid surfaces than in liquid media. Theoretically, the storage in plasmids of genetic information relevant to the breakdown of unusual environmental substances would be an efficient ecological and evolutionary mechanism. For most of the time, most of the microbes in a population would not squander their resources by assembling unwanted enzymes or even the plasmids that direct their synthesis. But when a new exploitable substrate appeared, transmission of plasmids throughout the population from the minority of bacteria carrying them would rapidly provide other organisms with the necessary biochemical machinery.

Though not yet fully proven, it may well be that such a mechanism explains some of the amazing catholicism shown by the scavenging microbes. Dr Waid's recommendation for exploiting the mechanism in pollution control is intriguing. He advocates 'the addition of small and sub-toxic quantities of biodegradable materials to the environment to prepare the way for the addition, deliberate or accidental, of larger quantities of such materials.' In other words, we should mobilize the microbial population of the soil by priming its citizens with small amounts of substances they are likely to meet in future. It is doubtless a feasible idea. But a far greater opportunity exists to harness scavenging talents more efficiently 'in house', in sewage treatment and the processing of specialized effluents, rather than in the wild. The possibility of tailor-making microbial strains for this purpose rather than (as at present) simply waiting to see what turns up, must be enormous.

Degradation and persistence

It has become customary to divide materials into 'bio-degradable' and 'non-degradable', the former being substances that, when discarded into the soil, the rivers, or the oceans, are broken down by microbial action. The latter group, including the earlier artificial detergents and such persistent insecticides as DDT, do not disappear but accumulate and thus cause untold troubles. The distinction is not absolute, however, because virtually all materials are degraded, either immediately or slowly over a period of time. One of the most industrious soil scavengers is *Cytophaga*, a bacterium that decomposes cellulose when dead plant tissue falls on to the soil. First discovered by Winogradsky, *Cytophaga* is unusual in moving by gliding — a primitive form of muscular activity based on flexing of the cell wall — and by being intensely specialized. Though several different species exist in the soil, in fresh water and in the sea, cellulose is the only food they can use for growth and energy. But they are voracious consumers. A spectacular demonstration of this is to lay a sheet of filter paper on a dish of agar and then inoculate it with *Cytophaga*. Orienting themselves in battle order along the cellulose fibres. the bacteria destroy the paper totally.

Other materials in the soil are broken down less rapidly and dramatically, but nonetheless inexorably. In one experiment, rye grass was grown in the presence of carbon dioxide in which the carbon was a radioactive isotope. The tissues of the grass thus contained 'labelled' carbon which could be detected by a Geiger counter. Some of the dead grass was added to soil, and six months later sixty per cent of the carbon had left the soil — after being released from the grass by microbial activity and thus converted back to carbon dioxide. But twenty per cent remained in the soil after four years. Some fibrous material persists for considerable periods, being incorporated into humus. Peat is also produced by microbial decay of plant debris. Subject to compression, the peat formed from the fern-like plants of the carboniferous era, 300 million years ago, in turn yielded coal.

Humus and coal have, of course, useful, natural functions. The persistent materials that cause concern are those with no value whatever. The scale of the problem can be gauged from the DDT saga. DDT has probably saved more lives and prevented more illness than any other substance ever invented by man, but we deployed it with criminal abandon during the 1950s and early 1960s, when we liberated some two billion kilograms into the environment. Much of this persisted either as itself or as an even more stable breakdown product. Not surprisingly, Alaskan Eskimoes and other peoples of the world who had never used DDT, or knowingly been exposed to it, began to accumulate the insecticide in their body fat. For this reason, and because of its disastrous effects on the breeding of certain birds and on other forms of life, DDT usage has been curbed drastically. Despite its undoubted short-term benefits, we broadcast DDT with such foolishness and crudity that we created a situation potentially much more serious than the hazards we were seeking to control.

A little thought would have helped. While we now recognize that biological pest control using microbes is far preferable to chemical overkill, it is also clear that we can easily create chemical insecticides that soil microbes will degrade speedily and totally. Chemists can incorporate 'handles' on to insecticide molecules, making them vulnerable to attack by enzymes produced by soil scavengers. The World Health Organization has already announced that it will consider for malaria control only new compounds that are degradable in the environment.

Pesticide-licensing authorities in many countries are beginning to insist on the same safeguard. Insecticides should be degradable within two or three months, which is sufficient to prevent them from building up in the environment.

Dr Martin Alexander is the microbiologist who first enunciated, in the mid-1960s, what he called the illusion of microbial infallibility. If a chemical is capable of being oxidized at all, the principle states, then somewhere on earth one can find microbes capable of catalysing its degradation. It is a notion that, the closer one looks at microbial ingenuity, becomes continually more persuasive. Apparent refractoriness may reflect only our own ignorance. Even at the time Alexander was writing, students were being taught that one common substance, benzoic acid, could not be broken down by anaerobic bacteria. There were cogent logical grounds and abundant practical evidence that this was so. Today we know of at least two strains of bacteria that, using contrasting methods, degrade benzoic acid anaerobically. The same story could be retold many times over. We are only just beginning to glimpse the full range of microbial dexterity in nature, the powerful and infinitely varied armoury of microbial enzymes capable of decomposing whatever we put in front of them.

Alas, it is just this knowledge that can easily encourage us — particularly in industry and agriculture — to believe that the microbes will always automatically cleanse the environment of our effete materials and our technological filth. That is a dangerous assumption. It would be infinitely wiser to recycle materials where possible, harnessing micro-organisms to help us do so; to 'build in' degradability into substances that have to be let loose in the environment; and to learn much more about the biology of the scavengers. We should work with the cleansing microbes, rather than trusting blindly in their dedication — as a vomiting drunkard assumes someone will mop up after him. Greater understanding of the scavengers would be invaluable. Chemical and biological evolution have occupied a time span of a few billion years. Chemists have been synthesizing novel compounds to a significant extent for barely a century or so. Only during the past few decades have we been casting masses of chemicals into the environment without thought for the morrow. There has probably been insufficient time, therefore, for the microbes to evolve machinery to deal exhaustively with all of these new challenges. The pseudomonads in particular (bacteria of the *Pseudomonas* genus) possess impressive batteries of enzymes capable of breaking down a wide range of substrates. Laboratory experiments, in which the pace of evolution has been forced in a test-tube, have shown how these enzymes can become adapted to deal with materials other than their 'natural' substrate. This process presumably occurs in nature. We may even be able to encourage it.

The scope for applying such new knowledge in disposing of specific industrial wastes is also clear. Achievements to date, which have been based almost entirely on the selection of efficient natural scavengers, have been greatly reassuring. Many industries generate effluent-containing phenols — chemicals, the best known of which is carbolic acid, that are strong disinfectants. Even they are degradable. Sewage microbes deal with tiny amounts, and the discovery that the bacteria responsible attack the phenol molecule more voraciously when given oxygen has led to vigorous aeration methods that are now employed in making safe phenolic wastes. How much greater the potential must be for tailor-making microbes for such specialized work. All that we have done so far is to provide natural candidates with encouraging conditions under which to operate.

Torrey Canyon and after

Of all recent environmental insults, accidents like that involving the *Torrey Canyon* are the most memorable and ghastly. In March 1967, the oil tanker *Torrey Canyon* was wrecked on the Seven Stones reef fifteen miles off Land's End, and for about ten days afterwards leaked its cargo of over 100,000,000 kilograms of crude oil into the sea. Pollution on this scale was unprecedented, and during frantic emergency discussions about measures to forestall damage to beaches and marine life, many different solutions were suggested. What the British authorities chose to do was to spray some ten million kilograms of detergents on to the sea and beaches in an effort to disperse the oil. But detergents are themselves toxic to marine life — so much so that an official report had to be written a year after the *Torrey Canyon* incident devoted largely to the harmful effects of the detergents, not the oil, on marine plants and animals.

What the UK authorities ought to have done but did not do was to turn to the microbes for assistance. The French were wiser. Faced with large quantities of oil that drifted into the Bay of Biscay, they applied chalk whiting (*craie de Champagne*), which sank the oil within a few hours, carrying it to the bottom of the sea where microbial scavengers could deal with it. As was clear from an analysis by Dr Quentin Bone and Norman Holme of the Marine Biological Laboratory, Plymouth, British fears that such tactics would have disastrous effects on fishing were not borne out. Though some of the oil simply separated again, and floated back to the surface, the French were relatively successful in staunching the menace by exploiting microbial skills. The British, using strong-arm tactics and the products of chemical industry, were less so.

Since the *Torrey Canyon* affair, microbiologists have devised improved methods of encouraging microbes to degrade oil slicks. Dr Philip Townsley, at the University of British Columbia, has developed a powder that is heavy enough to sink oil and which contains nutrients that help to promote the proliferation of scavenging bacteria. He estimates that organisms capable of breaking down oil exist in all places where spills might be expected, but that some of them need nutrients other than oil if they are to grow and do their work. Application of his grey powder encourages bacteria to multiply at such a rate that they can eliminate a light oil slick within three days.

Bacteria are also now helping to clean up oil tankers. Under the 'load-on-top' system, tankers wash out their empty tanks at sea with salt water and retain the resulting mixture as ballast. Soviet scientists became concerned about the amount of oil in this ballast that was being discharged into the sea at the Sheskharis oil terminal, established at the Black Sea port of Novorossiisk in the early 1960s. Though some oil was recovered from the ballast through gravity traps, tremendous volumes of water containing ten milligrams of oil per litre was still being pumped into the sea. After trying out various bacteria capable of degrading oil, Soviet microbiologists isolated a hardy species, well suited to the task, that could live in oily seawater. In 1970 they reported that, cultivated in large settling tanks together with two seaweeds also able to grow in oily brine, it efficiently oxidized the oil.

Israeli scientists have gone even further. Towards the end of 1974 they announced that they had developed oil-eating strains of bacteria that not only cleanse oil from the ballast of oil tankers, but also convert it into edible protein useful as animal fodder. Tankers dump about twenty billion kilograms of oil into the world's oceans each year, so the scope for the new process is sizeable. Professor Eugene Rosenberg and Dr David Gutnick, of Tel Aviv University, first identified an *Arthrobacter* species that thrived on light crude oil from Iran. They then discovered a pseudomonad that can devour even heavy crude oil, and turn over half of it into cellular protein. It is perfectly happy too, growing in the salty water of the Red Sea. Even a supertanker could be cleaned out within two or three days with its aid, leaving a slurry containing a vast reservoir of protein.

Oil tankers have to be taken out of service every two or three years and the inside scrubbed by hand — a laborious task that can last a month or so and thus cost the owner dearly in lost revenue. Microbial cleansing is designed to be cheaper, quicker and less wasteful. With the ballast tanks still full of water, about a kilo of bacteria would be added, together with some urea as a cheap source of nitrogen. Air would then be pumped into the tanks, encouraging vigorous bacterial growth and turning the oil waste into protein. Professor Rosenberg calculates that, on ballast voyages back from Europe to the Mediterranean port of Ashkelon, Israel's tankers could supply the country with a million kilograms of microbial protein every day — sufficient to satisfy Israel's need for animal feedstuffs. The system is not yet in practical use, but the potential is obvious and the groundwork has been done. The star performer — the astounding pseudomonad — has already been cast in the leading role.

Tricks and specialties
One of the least predictable tasks yet devised for microbial scavengers is that of clearing mines of methane — firedamp, as it is known to miners. Research workers at the Moscow Institute of Mines have invented a method of removing the gas by drilling horizontal holes, several metres long, in the direction of proposed future cuttings, and then pumping in a liquid medium containing bacteria able to live on methane. Air is also pumped into the seam to encourage the scavenger, which breaks down the lethal methane to carbon dioxide and water. After laboratory development, the technique has been applied practically at the Sukhodolskaya pit in the Donbas coal basin. It worked in a matter of days, though the Institute recommends that for routine use the bacteria should be introduced into the coal seams about a year before mining is due to start.

Equally intriguing is the suggestion from Dr Bernard Brown and his colleagues at the University of Manchester, published in mid-1974, that microbes can be harnessed in disposing of plastic — by common consent one of the most persistent and non-degradable of man-made materials. At present, the best that can be done with plastic waste (the volume of which more than doubles each year in the United States, Europe and elsewhere) is incineration or tipping. It is flung into holes in the ground created when other materials are extracted, or it is burned — which simply reduces its volume by about tenfold. In other words, it is not disposed of at all. The long-term solution is collection and recycling, and secondly (not alternatively) the production of degradable plastics. Just as 'soft' detergents have replaced the hard types not susceptible to microbial attack, so work is progressing towards the manufacture of soft plastics amenable to breakdown by common soil microbes. But for the foreseeable future there will still be an uncomfortable glut of plastic to deal with. Dr Brown and his colleagues have found that several fungi and bacteria are capable of growing on the chemicals released when plastic waste is broken down initially to simpler materials by nitric acid. The plastic does not even have to be separated. Mixed rubbish similar to town waste has also been oxidized with nitric acid, and the ensuing products used to cultivate microbes.

Dr Brown may have something. If bio-degradable plastics are marketed, they will probably not be able to replace conventional plastics for all purposes. Recycling will help enormously. But there will remain a disposal problem — one that is scarcely eased by the current development of 'photo-degradable' plastics, which are vulnerable to sunlight. This property is unlikely to be convenient when the plastics are in use, and is not particularly helpful in disposal, unless the waste that is now dumped into holes in the ground is to be laid out in the sun awaiting disintegration. A microbial method of dealing specifically with plastic would be worth having. Here, as elsewhere, we would be prudent to exploit the diverse talents of microbes in diverse ways.

Disposal, an out-dated concept

It is, however, in seeing effluent disposal no longer as simply a destructive exercise but as a synthetic one that the future lies. An early example of this occurred during penicillin manufacture during the 1940s. The first penicillin-producing moulds to be discovered gave only meagre amounts of antibiotic. But by selecting more efficient strains, and by incorporating an otherwise waste material, corn steep liquor, in the medium used to grow the mould, yields were boosted several thousandfold. Many projects are now underway to exploit wastes for similar purposes — some of them schemes to turn effluent into food (chapter 6). Researchers at the University of Aston in Birmingham have invented techniques for converting waste from caramel factories into yeast and moulds rich in protein, and for fermenting sawdust to ethyl alcohol. Another group of microbiologists, at the Wolfson Laboratory for the Biology of Industry (at University College, Cardiff), have evolved a whole series of what they term 'urban farming' techniques to make food out of such wastes as lubricating oils and cutting fluids from engineering works, via microbial growth. The head of the laboratory, Professor David Hughes, believes that 'man will not survive unless he undertakes this type of food resource development'. Elsewhere, at the University of Kent, Canterbury, research has yielded a method of growing *Saccharomyces fragilis* in coconut water waste to produce excellent quality protein. Coconut waste is a major environmental pollutant in many tropical countries, but the unusual carbohydrates that it contains are ideal foods for this yeast.

At present, some of these effluents and wastes are treated by conventional disposal techniques, in which microbes do no more than render them inoffensive. We squander gargantuan volumes of nutrients in this way. Over twenty million kilograms of sugar are thrown down refinery drains in Britain each year. The discarded whey in the United States represents a reservoir of approximately 204,000,000 kilograms of sugar and 36,000,000 kilograms of protein annually. Overall, if all the available whey, molasses and sulphite liquor from the paper industry were used to grow microbes for food, about a twentieth of the world's protein needs could be met. Currently in the United States only about a quarter of sulphite liquor is exploited to grow food yeast. And in many parts of the world these waste but potentially productive effluents are not even subject to orthodox disposal — they are merely gross pollutants.

An indication of the way in which the disposal business could develop in the future is provided by a project being conducted, on behalf of the US Bureau of Solid Wastes Management, at the Thayer School of Engineering, Dartmouth College, Hanover, New Hampshire. Researchers there are working on a method of breaking down waste paper with acid, to provide sugars that can then be fermented by micro-organisms to yield alcohol, fodder yeast, organic acids for the food industry or other end products determined by the microbe used. Dr Andrew Porteous, who has worked at Dartmouth and is now at the Open University in Britain, has written glowingly about such systems, pointing out that the likely costs would be very similar to those of the alternative now used — incineration. In 1970 the Greater London Council commissioned an incineration plant at Edmonton at a cost of £ 7.64 million. It processes gargantuan loads of material each day, turning it to clinker, fly ash and other effluent — and thus reducing its weight by sixty-four per cent. The running costs for this incomplete but wasteful destruction are considerable. The arguments in favour of a more positive approach are overwhelming. Some fifty per cent by weight of the solid refuse from a major city consists of paper, cardboard and vegetable matter. Even with more collection and recycling, refuse will continue to include a vast amount of these materials, which could be turned to good purpose by microbial action. A plant built to do so, rather than just burn its raw material, would also make a profit.

Old habits die hard in the sewage and refuse business. But the amount and variety of serious research now going into the emerging technology of waste usage, rather than waste disposal, confirms that the tide has turned. In very few decades from now, our past profligacy with material resources, and our neglect of microbial eagerness in husbanding those resources, should seem very foolish indeed.

8 Microbes as Chemists

Lord, I fall upon my knees
And pray that all my syntheses
may no longer be inferior
to those conducted by bacteria.

Anon.

The organic chemists' prayer, like many other supplications to the almighty, stems from a recognition of inadequacy and inferiority. However clever with their testtubes, whatever their station in the scientific hierarchy, chemists know that microbes can do better. In economy, elegance and dexterity, chemical transformations directed by bacteria and fungi are far superior to those carried out laboriously in the research laboratory. Not only that, microbes are also mass producers of industrial chemicals, churning out colossal volumes of substances that chemical engineers simply cannot synthesize as cheaply and easily. The opportunities for expansion in both areas — microbial craftsmanship and bulk synthesis — have never been greater than they are today.

Alcohol re-examined
Typical of the current status of microbes in industry is their present and future potential for mass-producing alcohol for innumerable purposes other than human consumption. Ethyl alcohol acts as a solvent for a wide range of products, including paints, dyes, lacquers and oils. Some is used as fuel, while considerable quantities are required as a feedstock for synthesizing other industrial chemicals. Around the turn of the century, microbiologists devised processes for making alcohol from surplus grain, carbohydrate wastes such as molasses, and later sulphite liquor and other materials. The special strains of yeast

employed were remarkably hard-working, even by microbial standards. Those maintained for industrial alcohol production today are miniature miracles of strength and efficiency, turning sugar into alcohol at a positively alarming pace and with an overall conversion rate of over ninety-nine per cent. Even the carbon dioxide generated alongside alcohol during fermentation is usually tapped off and sold as 'dry ice'. The surplus yeast, of course, may be used as food or fodder.

For several decades microbial production of alcohol competed (with varying advantage and disadvantage in different countries) with chemical methods of manufacture. Chemically, most alcohol is made from natural gas or from ethylene, which comes from the refining of petroleum spirit. Countries such as India have continued to favour yeastly synthesis, but elsewhere the cheapness of petroleum meant that during the 1950s chemistry began inexorably to take over from fermentation. The annual production of industrial alcohol in the United States is now over four billion litres. Yeast was formerly the principal contributor, but towards the end of the 1960s its share of the total had fallen to about a quarter.

Then came the decision of the oil-exporting countries in 1973-4 to begin charging a horrifyingly realistic price for their commodity. Suddenly, the complex economic comparison between chemical and microbial synthesis of alcohol changed. Calculations in 1974 showed that the two processes were at least competitive once more, and this has stimulated widespread re-examination of fermentation. Speaking at the Octagon group's meeting at the University of Manchester in March 1974, Dr R. N. Greenshields even suggested that industrialists reconsider converting alcohol, produced by fermentation, into ethylene, for use in turn as a starting material for making plastics. And Dr Duncan Davies reported that his company, ICI, was already manufacturing a small quantity of polyethylene by just this method in India.

Now that the rules of the game have been changed,

many questions have to be explored before we can make
optimum use of the tremendous skill and power of yeasts
in generating alcohol. A key issue is what substrate to use
for growing the organisms. While much needlessly wasted
'waste' can be exploited as microbial nourishment, the
idea of cultivating starchy crops especially as food for
fermenting microbes also looks increasingly attractive.
This is a strategic situation ripe with opportunities for plotting between microbiologists and agricultural economists.
Another angle to be investigated is the feasibility of converting otherwise waste cellulose into glucose (p. 144),
and thence into alcohol as an energy source—thereby
saving millions of years, as compared with the burning
of fossil fuels.

The oil sheikhs' action has not only forced us to re-
examine alternative, microbial, sources of alcohol: it has
also prompted an urgent reappraisal of direct replace-
ments for petroleum as a feedstock and source of energy.
Hence the suggestion of mixing alcohol with petrol for use
as motor fuel. This is entirely feasible, brings greater
cleanliness and was indeed practised in several European
countries during the 1930s. Only the pressures of adversity have been necessary to make us consider the idea once
more. Writing in a chemistry textbook published in 1934,
the American chemist James Conant commented: 'If industrial ethyl alcohol becomes a substance which is used
by the community in very large quantities (which might
occur if the gasoline supply were exhausted), the production of alcohol from cheap forms of starch would become
an enormous industry of the greatest importance.' Just
over forty years later, we stand on the threshold of a development of that sort.

Microbial aid in times of adversity
One previous episode in which political pressures threw
man upon the generosity of the microbes arose from a
scarcity of glycerine, which is needed in making nitro-
glycerine and other explosives, cellophane, textiles, paints

and innumerable other products. Louis Pasteur noticed that his alcoholic fermentations always yielded a little glycerine. Then, a few years before the First World War, the German biochemist Carl Neuberg discovered a way of modifying the fermentation so that glycerine became the major product. This was of purely academic interest at the time, because glycerine could be made cheaply from vegetable oils and fats as a by-product of soap manufacture. But the situation changed rapidly when war broke out. The British naval blockade of central Europe curbed overseas supplies of oils drastically. This stimulated Germany to develop Neuberg's process on a grand scale, and by 1918 she was using it to produce a million kilos of glycerine a month. Thus a piece of 'pure' research, of interest initially only to biochemists, rapidly became of crucial importance to the war economy of a major world power. The process was later abandoned, but the background knowledge still exists, to be harnessed on the grand scale if necessary at any time.

Microbial manufacture of two key organic chemicals, acetone and butanol, is another major chapter of industrial history — but one with great relevance to our plight today. Several microbes synthesize these important solvents, which are widely used in the plastics and other industries. The most prolific producers are species of the anaerobic bacterium *Clostridium*. They are versatile organisms, breaking down sugar and creating a range of products, including not only butanol and acetone but also various other acids and alcohols. By choosing a particular strain and arranging conditions to suit, microbiologists can boost the yield of the most desired end-product.

Chaim Weizmann, a chemist who later became the first president of the state of Israel, was responsible for developing the microbiological production of acetone — one of the first large-scale microbiological processes ever to be operated. Weizmann was born in western Russia in 1874. Despite his great scientific aptitude, he later had to go abroad to study, because of the restricted quotas for

admission of Jewish students to Russian universities. He
obtained his doctorate at the University of Fribourg,
Switzerland, in 1900, and four years later came to Britain.
After staying briefly in London, he accepted a post at
Manchester University, at that time a major centre of research for the chemical industry. In the spring of 1915,
David Lloyd George, who was then Minister of Munitions,
learned of his name from C. P. Scott, editor of the then
Manchester Guardian. There was a desperate scarcity of
cordite, at a time of increasing need, and Lloyd George
had been seeking help in finding new sources of acetone,
the vital solvent used in its manufacture, which was in
acutely short supply. Weizmann and Lloyd George met
and got on extremely well. Weizmann returned to his
laboratory determined to work without respite at the task
of devising a prolific alternative to wood distillation, then
the routine but quite inadequate method of making
acetone.

In his war memoirs Lloyd George records Weizmann's
success:

> In a few weeks' time he came to me and said: 'The
> problem is solved'. After a prolonged study of the
> micro-flora existing on maize and other cereals, also
> of those occurring in the soil, he had succeeded in
> isolating an organism capable of transforming the
> starch of cereals, particularly that of maize, into a mix-
> ture of acetone and butyl alcohol (butanol) . . . . in quite
> a short time, working night and day as he had promis-
> ed, he had secured a culture which would enable us to
> get our acetone from maize . . . this discovery enabled
> us to produce very considerable quantities of this vital
> chemical.

When, as Lloyd George puts it, 'our difficulties were
solved through Dr Weizmann's genius', the politician
suggested to the chemist an appropriate honour in recognition of a unique contribution to the nation. Weizmann
declined any such idea, but raised instead the question,

close to his heart, of a homeland for the Jewish people. That was the origin of the Balfour declaration, and of the subsequent creation of the state of Israel in 1949. Characteristically, Weizmann did not forget the microbes. As president of Israel, he saw to it that a department of industrial microbiology was established in the new nation's first university.

In the early years, acetone was the crucial product from Weizmann's fermentation, butanol being a more or less useless contaminant. But there was an increasing industrial demand for butanol after the war, and this ensured the survival of the process. Lessons learned in running the fermentation, which used huge but extremely pure cultures of single strains of bacteria, also proved invaluable when antibiotic-producing microbes were first raised on an industrial scale.

As with alcohol, the pattern of acetone and butanol production since the war has been one of increasing manufacture by the petrochemicals industry at the expense of microbial fermentation. In 1945 two-thirds of the butanol made in the US came from bacteria. For nearly three decades afterwards, the proportion declined. Then came the oil crisis, which has stimulated an urgent re-examination of microbial synthesis. It seems inevitable that this and many other fermentations will now begin to play an increasingly important role once more. The technology already exists, and there is also added impetus behind efforts to boost productivity — low yield was the only weakness of Weizmann's acetone-butanol fermentation.

Also currently the subject of phrenetic reassessment is microbial manufacture of butanediol, a key compound used in making synthetic rubber. Again, the historical precedents and lessons are clear. Many countries were cut off from supplies of natural rubber in Malaya during the last war, and synthetic rubber production boomed accordingly. The two principal raw materials can be obtained from petroleum, but the pressures of demand were so great that microbiologists investigated alternative sources. Research at the US Department of Agriculture laboratories at Peoria, Illinois, and at the National Research Council of Canada, Ottawa, showed that the more important starting material could be made from butanediol, a closely related substance produced by microbes. One bacterium employed was *Bacillus polymyxa*, which can break down starch in grain to glucose and then convert that to butanediol. Other methods use different microbes, growing on sugars.

Despite the dramatic progress that was quickly made in developing processes for making synthetic rubber in this way, interest waned after the war. Here too, petrochemicals proved to be a cheaper source of raw materials. But many microbiologists have continued to believe that the butanediol fermentation above all others would some day find widespread industrial application. That day may be about to dawn.

Acids and nutrient additives
One synthetic function for which microbes have always held the centre of the stage, despite competition from chemical alternatives, is in the manufacture of organic acids. Other than acetic acid, the main constituent of vinegar, these include citric acid, gluconic acid, itaconic acid, and many more that are important in the food, plastics and other industries. One of the earliest to be made in bulk by a microbiological method was citric acid, widely used throughout industry, particularly as an acidulent in foods and pharmaceuticals. Until the 1920s, when US Department of Agriculture scientists learned that the mould *Aspergillus niger* could produce it in vast quantities, most of the world's supply of citric acid had originated in Italy, where it was extracted from lemons. But manufacture by moulds proved much cheaper. Today, when western Europe alone makes sixty million kilograms of citric acid every year, the bulk comes from *Aspergillus* and its close relatives. Most producers grow the moulds on the surface

of nutrient medium in shallow trays. Others use submerged fermentation techniques, devised by the National Research Council of Canada and by the French company, Usines de Melle. Cane molasses is often the starting material, though Pfizer, by far the world's biggest manufacturer of citric acid, based its most recently opened plant in Ireland on beet molasses. Citric acid is an intermediary molecule in normal cell metabolism. The trick exploited in industrial manufacture is to add a chemical to the culture that allows the cells to accumulate citric acid, but prevents them from metabolizing it further. One such technique, widely employed today, was learned from a German company visited after the war by the British Intelligence Objectives Subcommittee.

Itaconic acid is a microbial product with many applications in the plastics and paints industries. It is used to make adhesives, paints, fibres and surface coatings, and is most efficiently synthesized by moulds such as *Aspergillus* rather than by chemical methods. Another, whose manufacture by microbes is an important, expanding industry, is gluconic acid. Produced by various moulds and bacteria, its applications range from inclusion in pharmaceuticals as a 'vehicle' for administering metals such as iron for anaemia treatment, to the prevention of scum in automatic washing machines. Finally, lactic acid (the acid responsible for souring of milk) is employed to preserve foods, to finish silk rayon fabrics, to make acrylic resins, to de-lime hides in the leather industry and for countless other purposes. 'Salt-licks' for cows, to stop milk production from falling away towards the end of winter, contain salts of lactic acid, which are also used to boost egg-production in chickens, to make polymers used in lacquers and varnishes, to promote electro-plating and to regulate the rate at which baking powders generate carbon dioxide. Industrially, lactic acid is usually manufactured by bacteria. A plant was opened in France a few years ago to make it from petroleum — but that now seems not a particularly bright idea.

Another group of acids produced in bulk by fungi and bacteria are amino acids, which are added to breakfast cereals and other foodstuffs as dietary supplements. The *rationale* behind this is to compensate for relative deficiencies of certain amino acids in particular proteins. Lysine, for example, often occurs in grain in proportions that are inadequate for human nutrition. Such imbalances (which were the basis for the distinction which used to be drawn between first- and second-class proteins) are now made good by including the deficient ingredients in pure form (or, of course, by mixing microbial and other types of protein together).

Amino acids can be produced by chemical methods, but research by Dr Shukuo Kinoshita of the Kyowa Hakko Kogyo Company, Tokyo, and other Japanese microbiologists, has demonstrated the blatant superiority of bacterial synthesis. All microbial cells contain a pool of amino acids, which they use to assemble their own proteins. But some of them also secrete amino acids through the cell membrane and into the liquid in which they are growing. The harvesting of nutrients released in this way is now the basis of worldwide production of amino acids for use as nutritional additives.

The first amino acid to be manufactured commercially by bacteria was glutamic acid. It is widely employed (as monosodium glutamate) not only as a dietary supplement but also as a flavouring agent. Industry formerly made it by breaking down plant protein, but in 1956 Dr Kinoshita isolated a bacterium that was far more efficient. Since then, his process has replaced the older one, while culinary demand for glutamic acid has also increased fantastically. US production is now some ten million kilos a year.

Today, knowledge of internal cellular control is proving invaluable in boosting yields of this and other commercially made amino acids. Mechanisms that normally regulate and harmonize the cell's internal machinery, preventing imbalanced production of vital metabolites, can be exploited when we wish to increase synthesis of a desired chemical. Mutants, natural or artificial, unable to curb their productivity are sometimes used. One organism is so freakish that a moderately sized industrial plant using it could turn out some 125 million kilos of monosodium glutamate per year. Alternatively, we can persuade otherwise well-adjusted bacteria to secrete gargantuan quantities of this popular additive by subtly altering the composition of the culture fluid. Two such methods are to deny the organism biotin (one of the B-group vitamins) and to include penicillin in the medium. The result is that the normally secure cell membrane becomes 'leaky', releasing the valued glutamate. It is gambits of this sort, combined with the prolific synthetic capacities of microbes, that render human chemists virtually irrelevant for making even such simple molecules as lysine and glutamic acid.

The second major type of dietary supplements generated in bulk by microbes are the vitamins. They are important not only to human health and wellbeing, but also commercially — in ensuring optimum egg and milk production from livestock, for example. Vitamins too can be synthesized chemically, but the prodigious quantities churned out by many yeasts and bacteria have made microbes the choice producers of several vitamins in industry. Two members of the B-complex, riboflavin and vitamin B_{12} (whose absence causes pernicious anaemia), are of outstanding significance. Microbial synthesis is the only source of vitamin B_{12} industrially — or anywhere else for that matter. Some of the industrial product is given to treat pernicious anaemia. Most is incorporated into poultry and other animal feeds. Vitamin B_{12} helps pigs and poultry make better use of vegetable protein, and it also improves the hatching rate of chicken eggs. Riboflavin goes into bread, flour and cereal products for human consumption, and also into livestock feedstuffs. It too promotes egg hatchability — and egg production. Another nutrient manufactured microbiologically (after many years of artificial synthesis) is beta-carotene, which animals including man convert into vitamin A. This vitamin is essential as a component of a vital light-sensitive pigment in the retina; deficiency causes night blindness. As well as being included as a nutrient *per se* in food and feeds, beta-carotene has proved popular as a colouring agent in foods. The chief microbial producers are two yeast-like moulds, *Eremothecium ashbyii* and *Ashbya gossypii*, whose faulty internal control systems are thought to be responsible for their unusually handsome productivity.

Antibiotics — the life-savers
Despite the variety of its other products, the 'fermentation industry' is today dominated by the £1,000 million per annum antibiotics market. Though a few totally synthetic types have proved useful, the overwhelmingly greater proportion of drugs given to combat occasional pathogenic microbes are themselves microbial products. Penicillin is the most famous, and the one whose dramatic benefits triggered off a worldwide search for other antibiotics. Alexander Fleming first discovered it, by capricious accident, in 1929, when he realized that something produced by a *Penicillium* mould growing on a discarded culture plate had inhibited the proliferation of nearby bacteria. Mass production of penicillin was later brought to fruition by Ernst Chain and Howard Florey under the exigencies of war. Heralded as a wonder cure, penicillin quickly vanquished the spectre of pneumonia and other bacterial conditions that used to threaten life. It remains one of the most widely prescribed and effective drugs administered to treat infection (though it and other antibiotics are frequently deployed indiscriminately and without good cause). Indeed, the only significant recent innovation by this sector of the pharmaceutical industry has been the chemical modification of penicillin to create 'new penicillins', rather than the introduction of novel man-made anti-microbial drugs.

Further antibiotics discovered in a phrenetic burst of activity in the late 1940s and early 1950s, following the dazzling triumphs of penicillin therapy, led to what Professor Ronald Hare has termed 'the metamorphosis of medicine'. Together with vaccines (which also exploit microbes themselves — in this case harmless, attenuated strains or skilled cells that create immunity but not disease), antibiotics have revolutionized the handling of infectious diseases by hospitals and medical practitioners. (Some have also proved useful in dealing with plant diseases, while others have been employed to treat cancer.) The extent to which this success story is a lasting one depends greatly on the prudence with which we continue to apply antibiotic drugs (p. 210).

A key figure leading the gold rush for the early anti-biotics was an exiled Ukranian, Selman Waksman, who spent most of his career at Rutgers College, New Brunswick, Maine. The discoverer of antibiotics such as streptomycin (highly effective against tuberculosis and other conditions), neomycin and actinomycin, Waksman was a soil microbiologist whose great achievements followed from his adherence to the principles advocated by Winogradsky and Beijerinck, which had been neglected with the passage of time. Waksman met the two pioneers on a grand tour of Europe in 1924, and went back to his laboratory at Rutgers fired with the need to adopt their comprehensive, ecological view of the soil's microbial population. He became fascinated by the symbiotic and antagonistic inter-relations between different organisms. This not only precipitated the hunt for antibiotics, but also spawned important developments in other practical fields such as agriculture. Waksman was thus led to his rich harvest of antibiotics by biological reasoning quite distinct from Fleming's fortuitous discovery of penicillin. All pathogenic organisms, he argued, sooner or later find their way into the soil. They then tend to disappear rapidly — possibly, he guessed, because they are destroyed by soil micro-organisms. If so, 'to what extent could such microbes be utilised for the purpose of producing chemical substances which would have a similar effect upon the pathogenic germs, in culture or even in the human body?'

Today we know the answer to that question. The vast majority of antibiotics now in use come from the selection and mass cultivation of soil bacteria and fungi. They can all be synthesized artificially in the chemistry laboratory, but microbial production is incomparably more efficient. Since 1945, over a thousand different antibiotics have been isolated, about fifty of which are today widely prescribed in human and veterinary medicine. Many of them are made by bacteria isolated during the days when explorers, missionaries, airline pilots and other travellers were being exhorted to bring home soil samples from all corners of the globe to be screened for antibiotic-producing microbes. As well as the better known ones—such as chloramphenicol, the penicillins and the 'broad-spectrum' tetracyclines —many others are reserved for specialized purposes. Their names commemorate such details as the area from which the soil microbe was isolated (lincomycin from Lincoln, Nebraska), the laboratory of origin (nystatin, discovered at the New York State Board of Health), patients' names (bacitracin, named after Margaret Tracy), and relatives of the discoverer (helenine, named after the wife of Robert Shope).

Miscellaneous manufactures

Such is the versatility of the microbes as chemists that, apart from the main product ranges for which they are responsible, they also provide us with innumerable specialized materials. Many bacteria, for example, produce excellent gums, whose applications vary from seed coatings to enamel paints. One type, dextrans, are widely used to 'extend' blood plasma for transfusion. They are also incorporated into surgical sponges, suture material, X-ray contrast agents, toothpastes and ice-cream cones.

Typical of the resourcefulness of the microbes is their recent emergence as producers of alginates, sticky substances employed as thickeners in ice-cream, soups and other foods. Alginates have long been extracted from seaweeds, but the job of collecting seaweed — formerly a minor industry in Scotland and elsewhere — is less popular these days than it used to be. This has led to an alginate shortage. A fermentation process for making alginates, announced in 1974 by the British company Tate and Lyle, should help to ease the problem.

Microbial products for which we are totally dependent on fungi as manufacturers include the ergot alkaloids, which doctors prescribe to treat menstrual disorders, intestinal bleeding and other conditions. Even perfumery is indebted to microbial ingenuity. Ustilagic acid, made by the fungus *Ustilago zeae*, is an important commercial precursor of musks, used in perfumes. The musks are very costly to prepare by alternative chemical methods.

The cornucopia of materials vested in the microbes of the world represents a unique resource. Man will no doubt extract many other useful chemicals from this natural store in years to come. The unparalleled variety of terrestrial micro-organisms, whose differing structures reflect differing composition, and the luxuriousness of their chemistry, manifest and latent, mean that the planet's microbial life is a vast and ubiquitous fund of synthetic skill, infinitely richer than that of any human chemist or research department. This is one important reason why we should seek to conserve all microbes rather than allow them to disappear through malice or neglect.

Transformation scenes
Since just before the Second World War, human chemists have been using microbial chemists in a rather more subtle fashion than simply as mass producers of acids, vitamins, dextrans and the rest. They have learned to exploit their micro-employees also as 'living reagents', to tinker delicately with the structure of molecules. Bacteria and fungi especially can conduct instantly and with consummate ease chemical reactions that would otherwise require several different steps and/or prove costly and time-consuming. Their major industrial role so far has been

in helping to manufacture steroids, including oral contraceptives and those like cortisone which are used to relieve arthritic conditions.

Dr L. Mamoli and Dr A. Vercellone, in 1937, were the first to realize that fermenting yeasts could alter the structure of steroids added to the culture medium. At a stroke, they had discovered a practical system for synthesizing the male sex hormone, testosterone. Over the next few years, further papers were published on steroid transformations, but none with similar practical significance. Then in 1949 came one of the most dramatic episodes in the history of medicine. Dr Philip Hench, working at the Mayo Clinic, Minnesota, found that patients suffering from chronic rheumatoid arthritis, given small doses of cortisone, improved miraculously after only a few days. Fever and pain subsided, swollen joints became almost normal, and limbs that had formerly been immovable loosened and were mobile once more. The lame walked. Alas, the high drama was short-lived, because continuous administration of cortisone gives decreasing benefits and also causes undesirable side effects. Nonetheless, cortisone and related compounds have proved immensely valuable in managing inflammatory diseases, and they are now major products of the pharmaceutical industry. The question in 1949 was: could microbes be persuaded to make cortisone from other cheap and widely available steroids?

The problem was tackled by Dr Durey H. Peterson and his colleagues at the Upjohn Company, Kalamazoo, Michigan. The incentives were clear. Against a background of increasing professional and public demand for the new 'wonder drug', existing methods of making it were pathetically inadequate. The choice lay between extracting cortisone from the adrenal glands of cattle (at least six thousand animals being needed to give as little as a hundred milligrams of the hormone) or synthesizing it chemically from bile acid by a laborious route involving thirty-two different chemical reactions and an overall yield of only 0.15 per cent. One particular step in the synthesis — introducing an oxygen atom at a specific point on the cortisone molecule — had proved extremely difficult. Dr Peterson's key discovery was that some fungi can execute this reaction speedily and without fuss.

Several plants — notably agave, Mexican yams and later soya beans — turned out to be rich sources of steroids which the fungi could convert into cortisone and related therapeutic agents. The Upjohn work not only caused a crashing fall in the price of cortisone; it also generated a surge of interest in the use of microbes as chemical converters, and led to the manufacture of other pharmaceuticals in this way. Two such are prednisone and betamethasone, which have greater anti-arthritic activity than cortisone itself. Prednisone, created by a relative of the diphtheria bacillus, was developed by researchers at the Schering Corporation, New Jersey. As well as being up to five times more effective than cortisone, it does not cause some of the troublesome side effects associated with the parent drug. Other medicines whose activities have been improved by microbial modification include alkaloids and tetracycline antibiotics. The chemical skills of micro-organisms have also been harnessed in making oral contraceptives. As recently as September 1974 the Mitsubishi Chemical Industries Company in Japan announced a new microbial process for converting cholesterol — from wool grease and fish oil — into the contraceptive steroid, norethisterone. Supplies of Mexican yams, source of the starting material for making the Pill, have been running short in recent years owing to the fast-increasing demand for oral contraceptives. Mitsubishi have wisely turned to microbial ingenuity to solve the problem.

Engineering with enzymes
The tools that microbes wield in achieving their multifarious transformations and syntheses are, of course, enzymes. Just as we have long appreciated the capacity

of microbes to make things, so industry over many decades has extracted enzymes from fungi, yeasts and bacteria, and harnessed them in its products and processes. Microbial enzymes have supplanted acids as a means of removing starch during textile de-sizing, and are now replacing malt enzymes during brewing. They are used to tenderize meat, to 'bate' hides in the leather industry, to modify the starches in breakfast cereals and to clarify fruit juices. One of their most appreciated but little known uses is in making soft-centred chocolates. Invertase (an enzyme obtained from yeast) is incorporated together with flavouring agents into sucrose, which is then covered in chocolate. Although the sucrose is initially solid, the enzyme converts some of it into the more soluble glucose and fructose. That is how chocolates acquire their liquid centres.

Microbial enzymes commend themselves in many ways as catalysts, compared with their chemical alternatives — particularly for use in the food industry. They are highly specific, and operate most efficiently under moderate conditions, requiring neither high temperatures nor extreme acidity or alkalinity. They are non-toxic and work in miraculously low concentrations, and their action can be speedily arrested at the chosen moment. These factors add up to a reassuring degree of safety, and also mean that the man in charge can control the process with delicacy and precision. Recently, however, biochemists have devised an even better way of exploiting microbial enzymes. Instead of being discarded after a single process, enzymes can now be immobilized on solid supports such as glass beads, and used many times over. Immobilization makes recovery and re-use easy, as well as stabilizing some enzymes that would otherwise lose activity. A related development is to entrap enzymes inside tiny semipermeable capsules. These form miniature artificial cells. The enzymes work away inside, converting substances that diffuse in from the outside as the products pass out the other way.

These new departures in microbial exploitation have stimulated the emergence of the new discipline of 'enzyme engineering'. Its arrival was marked in 1974 by publication of a report from the International Federation of Institutes of Advanced Study (IFIAS), which chose this subject as its first major study topic after its inception in 1972. 'The potential of enzyme engineering for generating production methods that save energy and materials, recycling and detoxification techniques that clean up the environment, and diagnostic and therapeutic devices that improve and extend medical care, is indeed impressive,' wrote Professor Carl-Göran Hedén, introducing the report. Already, the world sales volume for enzymes is some 100 million dollars, and rising at over five per cent per annum. That trend seems destined to accelerate with increasing incursions of enzyme-engineering into agriculture, health care and the food and other industries.

Indicative of the philosophy behind the new discipline is a study commissioned by IFIAS to assess the potential impact of enzyme-engineering in defining the optimal use that a less developed country could make of its photosynthetic energy. Suggested by Dr Gustavo Viniegra of the Universidad Nacional Autonoma, Mexico City, the plan stemmed from a recent and novel application of enzyme-engineering in the US. Mexico exports beef calves because she has insufficient feed to raise them to maturity, when they would fetch a higher price as exports. Mexico also sells considerable quantities of sugar abroad. But cornstarch producers in the US recently devised a method that uses an immobilized enzyme to create an alternative beverage sweetener, fructose, from cornstarch, which is in ample supply. Mexico might therefore be confronted with a surplus of sugar and a consequent loss of export income. The aim of the study is to calculate how best to use the solar energy that now goes into sugar cane production for export. One set of possibilities is to use microbial enzymes to convert both the cellulose in

the bagasse and the sugar into animal feed, or into glucose or alcohol for the chemical industry or for generating electricity. Other options include cultivating
alternative crops as feedstocks for raising microbes as a
source of enzymes for export or to create high-protein
nutritional supplements. Such is the versatility of microbes and their enzymes that the wide range of possible
interconversions requires considerable scrutiny before
a policy decision can be taken as to how best to harness
microbial skills in a situation of this sort.

In medicine the existing and potential applications of
microbial enzymes are legion. Some enzymes are already
prescribed as aids to digestion. Several are applied as
anti-inflammatory and debriding agents, and others are
injected to break down blood clots. Recently micro-
capsules containing enzymes have been used successfully
to treat patients suffering from conditions characterized
by a deficiency of vital enzymes in the body fluids.
Professor Thomas Chang, at McGill University, Montreal,
is developing these techniques, as well as encapsulated
enzyme devices that may be able to replace the artificial
kidney in removing urea and other wastes from the blood
of victims of chronic kidney failure — and of people
suffering from acute poisoning. One bacterial enzyme,
asparaginase, has been given to treat certain forms of
leukaemia.

As with industrial fermentations, enzyme synthesis can
be boosted by favourable environmental conditions, and
by genetic manipulation. Internal control mechanisms
usually ensure that microbial cells neither under- nor
over-produce enzymes. But some microbes are 'leaky',
others can be fooled into mammoth productivity, and
others can be tailor-made genetically to provide generous
amounts of wanted enzymes. One simple trick exploits
the fact that certain carbohydrates repress enzyme synthesis. Merely by excluding the appropriate substances,
we can persuade bacteria to assemble thousands of times
more molecules of enzyme than they would otherwise

produce. But the immaturity of enzyme-engineering as a scientific topic means that, as yet, less attention has been given to such techniques than to corresponding methods applied in conventional industrial microbiology. The potential is undoubtedly considerable.

Retting and wood preserving

The careful use of microbes and their enzymes to assist in making and preserving materials spans the centuries, from the empirical discoveries of primitive man to the latest breakthroughs of the research laboratory. Retting — the liberation of desired components of plant tissue through controlled microbial decomposition of other parts — is a venerable craft. In the oldest version, which has been used for several thousand years to make linen, bacteria release the bast fibres from flax and hemp. A 'cement' of pectin normally holds the cellulose fibres together, and the microbes attack this without at the same time destroying the cellulose. The process begins when the plant stems are immersed in water. Aerobic bacteria begin to grow, but they quickly use up all of the available oxygen. They are replaced by anaerobic organisms, which proliferate rapidly, attacking and loosening the fibres. Manufacturers of potato starch employ a similar retting technique to release the starch-containing cells from the pectin in which they are embedded.

Dr William Fogarty and his colleagues, microbiologists at University College, Dublin, have been pioneering a study that is a modern counterpart of the traditional art of retting. They began by trying to understand how bacterial action makes softwoods more permeable to treatment with preservatives. They found that, during storage in a fresh-water lake, the wood was attacked selectively by bacteria and thus rendered more porous. The microorganisms use their enzymes chiefly to break down pectins in the wood, though other related activities play a minor part. Dr Fogarty has found that the process can be promoted more efficiently in enclosed tanks inoculated with the most effective microbial specialists. One of these, *Bacillus subtilis*, produces a pectin-splitting enzyme that is remarkably stable—a great advantage both to the bacterium in nature and to its future commercial exploiters. The new technique is more effective because it conserves enzyme activity. It looks like having widespread use in preserving softwoods used by the building industry.

Microbial miners

'The use of microbes in mining and metallurgy is a relatively new idea, but potentially a profitable one', wrote Norman Le Roux, of the Warren Spring Laboratory, Stevenage, Hertfordshire, in September 1969. That was at a time when it was still just possible for us to believe that the resource crisis was a figment of the doomsters' imagination. Consequently, the notion of microbes extracting metals from low-grade ores and mining wastes seemed of little practical interest. It was a recondite stratagem, nothing more. Today, for obvious reasons, there is a regrowth of research into this intriguing aid to global housekeeping. A decade ago ores that either were inaccessible or contained relatively little metal could be dismissed as "unworkable" or 'uneconomic'. Today we cannot afford to be so choosy, and the idea of inviting bacteria to recover copper, nickel and even uranium from such rocks is a highly feasible proposition.

The oldest fossil bacteria yet known were found in strata up to two billion years old. It is likely, therefore, that microbes have always played a significant part in shaping geological change. Pyrites (metal sulphides), for example, are often leached naturally, yielding solutions rich in copper and other metals. One of the bacteria responsible, *Thiobacillus ferroxidans*, has been isolated from water flowing out of waste dumps in Britain containing copper suphide ore. The bacterium releases at least fifty parts per million of copper from the ore, liberating it into solution. *Thiobacillus* is an autotroph, which can oxidize iron, copper and other pyrites, solubilizing the metals they contain. The potential value of this leaching process in extracting metals from inaccessible rocks, and from ores and wastes poor in metals, is obvious. Over half the solid material produced in US copper mines contains insufficient copper to refine by conventional methods. A little of the metal is recovered by 'dump leaching', assisted by bacteria. Much more, normally discarded, is there for the asking.

Already, eleven different mines in the US are reported to be producing some copper by bacterial leaching. The technique has been practised since 1965 at Rum Jungle in Australia's Northern Territory, and in 1974 the Western Mining Corporation of Western Australia announced a microbiological method for winning another metal, nickel, from low-grade ores. The rock is sprayed with a dilute solution of sulphuric acid containing strains of *Thiobacillus*. Percolating through the ore, the liquid extracts high concentrations of nickel, which can then be recovered from the solution by electrolysis. Work is now in progress at the University of New South Wales, under Professor B. J. Ralph, to isolate organisms peculiarly well suited to leaching metals from other Australian ores. Rumania, Japan, Poland and Russia have similar active interests.

Uranium leaching is particularly promising, in view of the rapidly escalating cost of this metal and the ease with which bacteria scavenge it from the most unpromising materials. The Portuguese first reported that they had recovered uranium by microbial leaching in 1953; and at the Stanrock Uranium Mines in Canada, some eight thousand kilograms of uranium oxide have been leached monthly from rocks *in situ*. After the microbes have done their work on a mine face, the wall is hosed down and the water pumped to the surface. Though the solution contains only a low concentration of uranium the technique is viable because there is no need to haul up the mine shaft large quantities of rock containing feeble amounts of metal. Current investigations in South Africa, France and Sweden suggest that a significant fraction of uranium will be provided by microbial techniques in future.

Scientists at the US Bureau of Mines have employed microbes to recover manganese from insoluble manganese minerals. Their counterparts in Russia, at the Irkutsk Institute of Rare Metals, are studying 'the biometallurgical processes of dissolving and precipitating gold' — an obvious reference to microbial leaching. They have claimed 98 per cent extraction of gold from solution in fifteen to twenty hours and 'up to 30 per cent rock gold turned into solution'. Gold mining with microbes has also been tried out in West Africa. All such techniques have the disadvantage of being slower than orthodox mining; but financial pressures are conspiring to make that a consideration of diminishing importance.

Sulphur is another element whose microbial manufacture is being examined seriously in the light of modern economic circumstances. Microbiology textbooks have long recorded the fact that the anaerobic bacterium *Desulphovibrio desulphuricans* secures its energy by reducing sulphate salts to sulphur. Now a South African patent has been issued covering a process using bacteria to extract sulphur from gypsum. A second approach, being examined in several countries, is to release sulphur from the abundant sulphurous compounds in sewage by microbial action. Another prospect is to employ *Thiobacillus* to turn pyrites and other sulphides into sulphuric acid. As well as occurring naturally, great quantities of metallic sulphides are formed as waste products during coal gas manufacture. Only the price mechanism is needed to turn such otherwise obscure pieces of technical information as the biochemical habits of sulphur bacteria into practical techniques with major significance.

Bacterial power stations?

One of the most appealing of all prospects for exploiting microbes at a time of global crisis is that of harnessing them to generate power. At its simplest, this can be done by such crude tactics as growing algae and burning them or digesting them anaerobically to release methane, which is then burned in thermal power stations. This is certainly a serious concept, particularly in the tropics and subtropics. Far more sophisticated, and of potentially universal application, is the direct use of bacteria to generate electricity. M. C. Potter was the first scientist to demonstrate that this is possible, in a paper published in 1911. Earlier, William Grove had invented the prototype of the modern, conventional 'fuel cell', in which oxygen and hydrogen combine to make water—and produce an electric current in doing so. Potter found that bacteria and yeasts, growing in and fermenting organic matter, also generated weak electrical currents. Using a primitive 'microbial battery' of this sort, he achieved as much as 0.35 volts of electricity. Twenty years later an American scientist succeeded in extracting 35 volts, at a current of two milliamperes, from a microbial fuel cell.

Since that time, development of conventional fuel cells has proceeded apace, and though they have been used in manned space vehicles, they have never reached the considerable power outputs that have been predicted from time to time. Over the past decade, however, scientists have been enticed once more by the potentialities of biochemical fuel cells which tap electrical energy from processes such as the alcoholic fermentation of sugar by yeast. Recent years have seen patents taken out on such devices in both Britain and the US. Though still at the laboratory stage, microbial batteries could have a great future—not least because they can use as 'fuel' virtually any organic compound capable of being oxidized by microbes. And as we have seen, the list of such substances is virtually limitless. The major remaining challenge is still to understand precisely how bacteria and other microbes generate electricity. That knowledge could predicate considerable improvements in the efficiency of microbial fuel cells.

Limits to growth?

Power generation, alcoholic fermentation, microbial mining and other microbial labours are all open to several avenues of improvement. Merely screening many candidates and selecting the star performers often increases yields and efficiency many thousandfold. Altering the conditions of growth—by decreasing acidity, perhaps, or including particular nutrients or starting materials— sometimes helps enormously. So does the breeding of novel strains particularly skilled at the task in hand. The stupendous growth rate of microbes means that they are uniquely susceptible to rapid hereditary change. In future, the emerging science of genetic engineering will also be brought to bear on this problem. By increasing the number of genes in an organism responsible for making a product such as an antibiotic or hormone, an 'amplifier effect' should be possible, such that productivity is boosted astronomically. One American company, the Cetus Corporation of Berkeley, California, has recently been established with this one aim: to screen vast numbers of micro-organisms and to improve their genetic capacities for chemical synthesis and other tasks.

Much of this chapter has hinted at an uneasy comparison, highly unfavourable to human chemists, between their skills and the creative talents of the invisible microbes. One chemist, the Englishman Edward Frankland, stands out however as one of the first of the few scientists to perceive the magnificence of the prize that this planet possesses in its microbial population. Writing in 1885, he gave this prediction:

> The position of micro-organisms in nature is only just beginning to be appreciated. Their study, both from chemical and biological viewpoints, is, however, of the highest importance to the welfare of mankind; and I venture to predict that, whilst there is no danger of their being spoiled by petting, or by their welfare being made the special care of sentimentalists, these lowly organisms will receive much more attention in the

future than they have in the past. Their study leads the enquirer right into those functions of life which are still shrouded in obscurity, and it is to be hoped that these investigations will not be retarded by mischievous legislation.

No such legislation has been necessary, and as we shall see in the next chapter the microbes have abundantly vindicated Frankland's vision in revealing the chemical and hereditary basis of life. But, approaching a century later, we could well echo his views about the still emerging recognition of our indebtedness to the microbial world. Not one but several confluent global crises have been necessary to drive the message home. The microbes themselves always react more speedily to environmental calamity.

9 Servants of Science

'The organism selected for our investigation was the K-12 strain of *Escherichia coli.*'

The number of research papers carrying those or similar words must long since have passed the million mark. *Escherichia coli* is one of the most comprehensively studied of all living creatures on earth. Scientists have scrutinized its structure, its chemical behaviour and its genetic machinery in such detail that they have recently been able to consider seriously writing a total description, a complete specification of every molecular constituent, of this popular bacterium. But researchers have not been drawn to *E. coli* only by its intrinsic fascination and lovability. They have found it, and some of its colleagues, perfect experimental material for investigating the 'secrets of life' — the complex chemical and genetic operations that underly the life not only of bacteria but of all animals and plants. Together with yeast, the mould *Neurospora crassa* and a handful of other micro-organisms, *E. coli* has given us a considerable understanding of the physical basis of life. That knowledge, provided by the microbes at an accelerating pace since around the turn of the century, is now a major component of the corpus of modern scientific theory, a fundamental under-pinning of physiology, botany, zoology and the other biological disciplines. But that is not all. It also has far-reaching implications for such applied sciences as agriculture and medicine. And in cultural and intellectual terms it is at one with Newton's mechanics, Darwin's evolution and Einstein's relativity in giving us insight into ourselves and the universe we inhabit.

The chemistry of life
Man's appreciation of these things began, as so many good things begin, with fermentation. A few years before

172

Pasteur discovered that yeast cells promoted alcoholic fermentation, another Frenchman, Joseph Gay-Lussac, worked out the chemical equation for the process. He found that he could express the production of alcohol and carbon dioxide from glucose as a strict quantitative relationship. So crystallized one of the earliest conflicts between chemical theories of organic change and 'vital' interpretations, according to which such transformations were inseparable from life itself. In the case of fermentation, the debate devolved upon whether sugar had to pass into living yeast cells to be transmuted into alcohol, or whether chemicals produced by the yeast could effect the change in the absence of living cells. We now know that the dispute was unnecessarily polarized. But it was the resolution of that argument that gave rise to the science of biochemistry.

Two German chemists, Eduard and Hans Buchner, working at Tübingen in 1897, provided the answer to the conflict over fermentation. Like Fleming's penicillin, their discovery was accidental. While preparing yeast extracts for medicinal purposes, they ground some brewers' yeast with sand, mixed it with *kieselgur* and squeezed out the juice. Then, in an attempt to preserve the juice and prevent it from going mouldy, they added large quantities of cane sugar. (Concentrated, rather than dilute, solutions of sugar tend not to support microbial life because the cell membranes become ruptured by osmosis.) The Buchner brothers were amazed to find that the sugar fermented rapidly. As no microbes were present, the yeast juice alone must have been responsible. Their announcement 'that the production of alcoholic fermentation does not require so complicated an apparatus as the yeast cell and that the fermentative power of yeast juice is due to the presence of a dissolved substance' was the key moment in launching a triumphant movement which, over the decades, has shown that not only fermentation but all other metabolic processes too are chemically determined.

What the Buchners had discovered was enzyme activity.

They named their 'dissolved substance' zymase, and a few years later the Englishman Sir Arthur Harden and his colleagues took the first steps in showing that this was in fact a complex mixture of different enzymes, each one of which catalyses a single stage in the stepwise degradation of sugar into alcohol. At this point a pattern began to emerge. Discoveries made initially using micro-organisms would, one after another, prove relevant to 'higher' plants and animals. Thus the great German biochemist Otto Meyerhof, some fifteen years after the work of the Buchner brothers, demonstrated that some of the same enzyme reactions of yeast fermentation also occur in animal muscle. Later named the Embden-Meyerhof-Parnas scheme, after the three scientists who chiefly unravelled its detail, the breakdown of glucose to alcohol entails a dozen or so individual stages. Most of them are common to all forms of life on earth, though the sequence of reactions ends in different ways in some species. The astounding range of different alcohols, acids and other products made by bacteria, for example, reflects specialized enzymes at the terminal point of the pathway. In animals, the end-product of fermentation is normally oxidized through the 'Krebs cycle' (p. 177), rather than released unaltered. If we run a twenty-six-mile marathon however, particularly if unused to such exertion, our muscles are liable to become depleted in oxygen and produce lactic acid as the end-product. At such moments, our muscles are behaving very like a fermenting micro-organism.

Elucidation of fermentation in chemical terms was the key event that established a new scientific discipline. Where there were chemists, biochemists now began to prosper. Even more important, the growth of biochemistry triggered a deepening recognition of the unity of all life on earth. Darwin had discerned the common pattern of evolution in plants and animals and had painted his great portrait of organic change. The natural historians, forerunners of today's ecologists, had long been aware of the underlying similarities of structure and behaviour within the rich profusion of life. But to demonstrate that we, yeast cells, bacteria, cabbages, caterpillars and antelopes use precisely the same chemical reactions to secure energy from sugar introduces an altogether more profound perception of unity. Realization of biochemical universality not only vindicates Darwin and implies a common origin for all life on earth: the new sense of oneness also has emotional significance, altering our relationships with, and attitudes towards, all other inhabitants of the planet.

The microbes have been able to give us this understanding by virtue of their smallness and sophistication. Their size, combined with their prolific growth rate, makes them ideal subjects for laboratory research. And the fact that microbial metabolism is every bit as sophisticated as our own — rather than primitive, as we once supposed — means that microbes have, time and time again, revealed metabolic manoeuvres that have later been detected in human and other animal cells. Yeasts, the star bacterium *E. coli* and other micro-organisms have proved invaluable as whole cells, as sources of individual enzymes for research and as living models for investigations into the ways in which various enzyme reactions are linked together into the complex web of cellular metabolism. Insights gained from research on microbes have frequently suggested corresponding probes into human metabolism. Conversely, biochemists have often turned to bacteria in particular as convenient research candidates that are so much easier to scrutinize than pieces of the human or animal body. Slices of liver, kidney and other animal organs are sometimes studied, but these tissues have the disadvantage of specializing in only certain types of activity. Microbes are much more self-sufficient and versatile.

One of the pioneers of microbial biochemistry, Marjory Stephenson, likened the biochemist's task to that of an observer trying to understand life inside a house by making a careful scrutiny of the food and supplies that arrive at the house, and by examining the contents of the dustbin. From these limited sorts of information, the observer tries to deduce what is going on behind the closed doors. On this analogy, one of the merits of microbes as research tools is that they allow the biochemist to open the doors (gently, partially or violently) to watch the domestic scene; to interfere selectively with family relationships and make deductions from what follows; and even to select particular rooms, such as the kitchen or toilet, for study in isolation.

Biochemists, can, ever so delicately, remove the microbial cell wall and membrane, and study life processes uncomplicated by the selective entry and exit barrier that normally regulates interchange with the outside world. They can investigate isolated pieces of cells in the same way. They can replace normal nutrients by abnormal ones (molecules tagged with radioactive atoms or modified in other ways) and then trace what happens to them — as the household observer might deliver pork sausages instead of beef and record whether the wrapping paper appeared in the refuse. They can even impose a block on specific important cell reactions. Some of the early biochemists did so by poisoning microbes with chemicals that trapped intermediary metabolites and thus arrested the functioning of key pathways. Nowadays, we know that the microbes themselves spawn occasional mutants with barriers of this sort, which are therefore ideal specimens for investigating metabolic pathways. Many features of cellular chemistry have been clarified by observing the effects of antibiotics — including that of streptomycin on ribosome function and that of penicillin on cell wall synthesis. Pharmacologists have long dreamed of designing antimicrobial drugs on the basis of their understanding of cell metabolism. In practice, the reverse has been far more common — the empirical discovery of antibiotics and their later use as tools in biochemical research.

One of the most crucial insights revealed by *E. coli* has been that of 'competitive inhibition'. This occurs when an enzyme is offered a substance whose structure closely (but not closely enough) resembles that of its substrate — the chemical it normally acts upon. The decoy molecule attaches to the enzyme in the usual way, but cannot be converted further. It thus inhibits the enzyme from doing its routine work. This technique has been widely employed as another selective tool for dissecting and charting metabolic reactions in the cell. Two particular applications have been outstandingly important. The stratagem was instrumental in allowing Sir Hans Krebs to uncover the cyclical series of reactions by which cells oxidize the end-product of fermentation and thus acquire energy. Like fermentation itself, the Krebs cycle operates in organisms ranging from algae to antirrhinums and alligators. It is universal in all aerobic forms of life. A second merit of the discovery of competitive inhibition was that it explained how the sulphonamide drugs like M and B work: they operate by imitating an essential growth factor and thus inhibiting the enzyme that normally handles it. Deprived of vital sustenance, the cells die. Since this discovery, other 'metabolic antagonists' have been developed and have been given to treat both infectious and malignant disease.

Microbes have not only played a masterly role in giving us our knowledge of the biochemical basis of life in animals. Algae also provided the key to our understanding of photosynthesis — the mechanism that green plants use to build up sugars, starches and other carbohydrates from carbon dioxide gas, using the sun's rays as energy. In 1946 Melvin Calvin set up a research group at the Lawrence Radiation Laboratory of the University of California whose objective was to trace the reactions behind this process. The idea was to expose a photosynthesizing plant for a time to carbon dioxide 'tagged' with radioactive carbon, and then to arrest metabolism and examine which substances in the leaves now contained the labelled carbon. By repeating this procedure at various times after exposure to the radioactive gas, Dr Calvin and his team hoped to trace the sequence of chemicals carrying the carbon isotope, and thus to plot the series of reactions through which the plant normally built carbon into carbohydrate.

Initial efforts using whole plants were disastrous, producing no discernible labelling pattern. The plants were photosynthesizing far too rapidly to allow the desired monitoring. Repeated attempts failed. So did tests in which the researchers exposed single isolated leaves to labelled carbon dioxide and then rapidly immersed them in alcohol to kill the cells and thus 'freeze' the pattern of labelling. The few seconds taken for the alcohol to penetrate into the cells were enough to obscure any clear sequence of carbon transfer. So Dr Calvin's group turned instead to algae — *Chlorella* and *Scenedesmus* — for most of their experiments. The original plan now worked beautifully. As a result, the research team had soon constructed the 'Calvin cycle', the reaction cycle that all green plants, including algae, use to fix carbon dioxide from the atmosphere into the carbohydrates that are consumed by animals. As well as being responsible for the greater part of the planet's photosynthesis, a phenomenon vital to life as we know it, the algae also showed us how it works.

Heredity and the DNA story

Like Calvin's later attempts to understand photosynthesis, the study of inheritance reached an impasse, in the 1920s and 30s, until microbes were used. The pea-breeding work of the Augustinian monk Gregor Mendel sixty years before had been rediscovered and extended. Geneticists were now compiling an impressive picture of how genes on the chromosomes govern heredity and account for variation in plants and animals. What was blatantly lacking, however, was any connection between this research and the developing science of biochemistry. Geneticists dealt only with organisms as they were. They counted how many members of a population had brown feathers or wrinkled seeds, and they monitored the eye colours of breeding groups of fruit flies. But they had no means of knowing how such visible characteristics were produced by the genes. Investigation along these lines virtually ground to a halt after a paper by Sewall Wright in 1917, in which he tried to relate coat colour in mammals to the enzymes that might be responsible. It was a well reasoned paper, far in advance of contemporary thought, but Wright simply could not pursue his theory any further.

Neurospora crassa provided the key to progress. This mould can be easily grown in the laboratory and its rapid asexual reproduction provides a plentiful supply of genetically identical individuals, which are ideal starting material for genetic studies. Most important, it is a perfect organism with which to demonstrate dependence on specific nutrients in the environment. In 1941 Dr Geoge Beadle and Dr Edward Tatum, working at Stanford, California, reported that after irradiating the mould with X-rays they had found occasional mutants that could not grow on the standard culture medium, but would thrive as usual if certain vitamins were included. Changes in hereditary constitution thus caused specific synthetic liabilities. What happened in a mutant, Beadle and Tatum suggested, was that a gene that normally directed production of a particular enzyme had been altered, creating a corresponding nutritional handicap. This 'one gene-one enzyme' hypothesis has proved to be substantially true and was crucial in forging the link between genetics and biochemistry, the foundation of modern biology. The hypothesis holds good not only for microbes but also for plants and animals.

As every schoolchild now knows, the substance that composes the genes, and thus acts as the vehicle for hereditary instructions passed down the generations, is DNA. The unfolding story of how DNA works—arguably the greatest chapter in the history of twentieth-century science

— has depended at every turning point on microbial aid. The saga began with a shy biochemist, Fred Griffith. In 1928 he demonstrated the phenomenon of *transformation*, in which a chemical (now known to be DNA) from one microbe alters permanently the behaviour of another strain. He proved his point by injecting mice with non-virulent pneumococci, together with (non-living) material from a virulent strain. The bacteria were transformed in virulence—and killed the mice. Griffith was a cautious man and, though he discussed with his colleagues his hunch that DNA was the active substance, he did not mention the possibility in his written report.

So it was that the train of events that was to lead to the Nobel prize for James Watson, Francis Crick and Maurice Wilkins in 1962 passed to the Rockefeller Institute in New York. There, in 1944, Oswald Avery and two colleagues proved conclusively that DNA was the chemical that transformed pneumococci. It is difficult to over-estimate the significance of this breakthrough. For the first time, a definable chemical compound had been shown to change the hereditary makeup of a living organism. Moreover, once it had played this role it was capable of reproducing itself 'in amounts far in excess of that originally added'. The number and variety of experiments carried out by Dr Avery's team — all on pneumococci — made their conclusion unambiguously certain.

The Rockefeller report signalled one of the most concerted research efforts ever known. Throughout the world, biologists began to cultivate bacteria with which to probe further the molecular basis of inheritance. A spate of discoveries followed. Dr Tatum moved to Yale University and from there in 1946 he and Dr Joshua Lederberg reported the occurrence of sexual transfer in *E. coli*. This was the first demonstration of sex in bacteria, highly significant because the mixing of genes during sexual transfer generates the new genetic combinations that are acted upon by natural selection—the motive force behind evolution. In the light of the rapid and abundant multiplication of bacteria, this revelation showed that they were even more valuable research tools than they seemed already. Dr Lederberg made full use of his discovery, and became a leading figure in the new science of microbial genetics. In 1952 he and Dr Norton Zinder reached another landmark when they showed that viruses too could transfer genetic material between different strains of *Salmonella typhimurium.*

Knowledge won by this increasing array of microbiological techniques was building up to two crucial events. The first was the announcement of the structure of DNA in 1953 by Watson and Crick, working at Cambridge. The significance of their work is contained in the last sentence of their paper, one of the biggest understatements ever to be published in the scientific literature: 'It has not escaped our notice that the specific pairing we have postulated immediately suggests a possible copying mechanism for the genetic material.' What Watson and Crick had done, of course, helped by the microbes, by Maurice Wilkins (their fellow investigator from King's College, London) and by other previous researchers, was to describe the DNA double helix and to show how the double-stranded molecule works. The two long strands carry coded hereditary instructions, the units on each being complementary, so that when DNA (and its parent cell) divides, the single strands can assemble new corresponding twins.

All that remained was the cracking of the code by which the sequences of units in the DNA genes determine the sequence of amino acids in enzymes and other proteins whose production they direct. Francis Crick, together with another Cambridge researcher Sydney Brenner, played a major role in this work. It ended successfully in 1967 when, by courtesy of *E. coli*, the final details were confirmed. Not the least exciting piece of knowledge to emerge from this research was that the code, too, is universal. The hereditary instructions that determine our hair colour, the capacity of our liver to cope with alcohol, our inborn

athletic potential and countless other characteristics reside
in the same substance and in the same coded form as they
do in sticklebacks, sunflowers and sewage bacteria.

Self-restraint among the cells
The second great achievement in modern biology, to
which many different microbes and microbial investigators contributed, crystallized around 1960, when the
Frenchmen François Jacob, Jacques Monod and André
Lwoff provided a coherent explanation of how living
cells regulate themselves. The question their work began
to answer first arose during the first half of this century.
As knowledge of biochemistry accumulated, biologists
became increasingly disturbed by the riddle of how the
complex machinery of living cells is regulated. Given an
adequate supply of food materials, what prevents cells
from running continuously at full throttle? Their physical
structure, particularly their selective membrane, imposes
obvious constraints, but there is evidence of extremely
delicate and flexible regulation of metabolism that cannot
be explained on this basis.

One clue came from experiments using mutant bacteria
with a blocked enzyme reaction. Normally, such a block
would cause the previous metabolite in the reaction sequence to accumulate. If the end-product of the pathway
was added in excess, however, the intermediary failed to
accumulate. What was happening was that the end-product
was inhibiting the action of previous enzymes in the pathway. In normal cells — microbial, human, animal, plant —
this ingenious feedback device prevents over-production
by some metabolic pathways. Like a governor on a steam
engine, feedback automatically retards the process if it
begins to exceed its ideal working rate.

But that is not the entire story. The discoveries that led
to the historic analysis by Jacob, Monod and Lwoff go back
to the beginning of the century, when several scientists
were struck by the adaptability of many microbes. One
discovery was that a bacterium produced a starch-splitting

enzyme when grown in medium including starch, but not when the medium lacked starch. Another was that some strains of yeast generated an enzyme capable of breaking down galactose when they were incubated with this sugar, but not when transferred to a medium containing glucose instead. Then, in 1936, Marjory Stephenson found that this enzyme activity appeared. in the presence of galactose, without the number of cells increasing. The yeast was not simply multiplying and spawning cells able to make the enzyme: the change occurred within existing cells.

We now know that many microbial enzymes are controlled by either *induction* or *repression*. Inducible enzymes are principally those that bacteria require occasionally to break down sugars and other foodstuffs. Some bacteria, for example, can utilize lactose, but they produce the requisite enzymes only when confronted with this particular sugar. The cells have the genetic capacity to make the enzymes to attack the sugar, which itself induces their synthesis. This economical arrangement means that bacteria do not squander their materials or energy in synthesizing enzymes when they are not required. Induction is often sequential. The sugar induces synthesis of the first enzyme in a pathway. This acts on the sugar, producing a metabolite that then induces the next enzyme, and so on.

Repression is the reverse of induction. Here enzyme synthesis is inhibited, usually by the end-product of the metabolic pathway. In contrast to inhibition of enzyme *activity*, repression of *synthesis* usually covers all of the enzymes in a pathway. The reactions affected are normally those that cells use to build up cell materials, rather than those that digest foodstuffs. Repression is another beautifully effective economy measure. As soon as any of the metabolic production lines in the cell begins to exceed its ideal productivity rate, every stage in the process is retarded simultaneously.

The achievement of Monod and his colleagues was to

offer a satisfying explanation of the genetic basis of metabolic control. They showed that it is the genes themselves
that are regulated. In addition to genes that determine
synthesis of enzymes and other proteins, there are
'regulator' genes whose job it is to switch the other sort
on and off. In the case of repression the genes responsible
for the enzymes of a particular metabolic pathway are
clustered together on the DNA strands. Alongside is the
regulator gene, which produces a repressor capable of
switching off the cluster of genes. A similar scheme controls enzyme induction. Since the Frenchmen first put
forward their interpretation (for which they received a
Nobel prize in 1965) it has been fully vindicated. Many
of the finer details have been worked out. Microbes,
principally bacteria and viruses, have played the major
part in this.

Like the biochemistry that was being elucidated half a
century ago, the emerging picture of metabolic regulation
has far wider relevance than just to micro-organisms.
Essentially similar mechanisms also control enzyme
production in human and other animal cells. They are
not only responsible for ordering metabolism in mature
cells. They also play a part in directing the development
of animals and plants from egg and seed to maturity.
Many different chemical reactions have to be brought into
play according to an accurate timetable during this unfolding programme. Interference with normal control
mechanisms may also at least partially explain the origin
of some forms of cancer — a disorder of growth in which
cells multiply wildly without restraint.

In microbiology itself, knowledge of metabolic regulation is of interest because organisms with faulty control
machinery can generate gargantuan quantities of enzymes
or of the products whose synthesis they catalyse. Mutants
of *E. coli* are known that synthesize up to seven per cent
of their total cell protein as a single enzyme. Such microbes
have obvious potential in industrial fermentations and
allied processes.

Vital repairs

DNA is by far the most important component of the living cell. As the site of genetic instructions, it is comparable with the information store of a large computer, holding the data upon which a highly automated factory depends at every moment. Cells may sustain severe insults to other parts of their structure and recover or at least survive. But even minor damage to DNA can prove fatal. A slight alteration in the coding at just one point along the enormously long DNA helix can block the production of a protein vital to survival or trigger the synthesis of an abnormal one with lethal potential. Like a small spanner thrown into the heart of a complex piece of machinery, the most short-lived and feeble burst of radiation can do mortal harm to the genetic material.

But we all manage to withstand some radiation — whether it is the continuous ultraviolet radiation from the sun, occasional exposure to X-rays or the radioactivity to which people living in certain geological regions are regularly subjected from underlying rocks. There is no such thing as a totally safe dose, and regulations to limit exposure in industry and elsewhere have been repeatedly improved in recent years. At the same time, we (and other living creatures) are not as vulnerable to radiation as one might suppose, considering the sensitivity of DNA and its pivotal importance in our cells. Why?

One of the tiniest of bacteria, *Micrococcus radiodurans*, has given us part of the answer. As its name suggests, this little coccus is unusually resistant to even heavy doses of radiation. Not for the first time, a micro-organism illustrative of one of the exotic extremes of existence has helped us understand creatures like ourselves that live an altogether more moderate life. Professor Howard Flanders and his co-investigators at the Yale University School of Medicine were the first to discover the secret behind the unrivalled hardiness of this microbe, and to realize that the skills it possesses in abundance also operate to a lesser degree in human cells.

The explanation is one of exquisite elegance. DNA is indeed vulnerable to radiation. If accidents could be repaired quickly, however, they need not prove disastrous. That is the technique that cells have acquired during evolution for dealing with a recurrent hazard. The moment any region of the double helix is damaged by radiation (or, for that matter, by certain harmful chemicals), a four-stage emergency procedure swings into action and restores it to a normal operational condition.

First, the cell detects the defect in its own DNA. Then two enzymes come into play to excise the damaged fragment. They have no other role in the cell but to cut out defective regions of DNA after an accident. This leaves a gap in one strand of the double helix. Another enzyme now assembles an identical copy of the missing piece of DNA as it was before being damaged, using as a template the coded instructions on the complementary strand. Finally, a fourth enzyme inserts and joins up the newly constructed components, so completing a fully functional double helix. Genetic continuity has been preserved.

M. radiodurans, which first revealed this drill, owes its outstanding radiation resistance to an incredibly efficient and well co-ordinated repair mechanism. It can withstand several different crises at the same time, and may even be able to restore double-stranded breaks — though how it does so has as yet escaped the attention of researchers. Other micro- and macro-organisms have similar, though less striking, abilities. In all creatures, it seems, the cells continuously monitor DNA and replace unsatisfactory regions. An ancillary mechanism employed to do so is 'photo-reactivation'. A special enzyme combines with a DNA region put out of action because of obstructive cross-links caused by radiation. Using light energy, the enzyme then separates from the DNA, magically restoring it to its original condition.

Unlike the replacement of damaged fragments, which can be done in darkness, photo-reactivation depends on exposure to the light. The enzyme, though first detected in bacteria, also occurs in fish, amphibia, reptiles and marsupials. The fact that other mammalian cells do not carry it suggests that this enzyme has some other principal function when it is present, and that its DNA repair work is an accidental benefit. The excision, re-assembly and insertion enzymes, on the other hand, may well be universal; human cells certainly contain them. Their 'survival value' is emphasized by the consequences that follow when they are missing or inoperative. The human disease Xeroderma pigmentosum is a genetically determined condition in which certain skin cells, unlike their counterparts in healthy individuals, cannot rectify damage to their DNA. The presence of unrepaired DNA leads to a fatal form of skin cancer. Thus an esoteric piece of biological knowledge provided by an obscure bacterium has proved invaluable in understanding a serious disease.

Evolution in a test-tube

One difficulty in grappling with much contemporary research in microbiology is the unusually heavy jargon indulged in by microbial geneticists in particular. Within a couple of decades, a topic that was readily accessible has become as opaque to the outsider — including scientists in other disciplines — as atomic physics. That is a pity. While this retrograde development can be partially excused on the same grounds that other specialists use to justify their outwardly incomprehensible deliberations, there are many developments that should be more widely reported but are not, because the language barrier is considered to be too formidable.

That was probably the case with an important paper published in 1973 by a research team headed by Dr John Campbell, of the School of Medicine, University of California, Los Angeles. What Dr Campbell and his colleagues had achieved, with a practical elegance that matches the conceptual beauty of their discovery, was to monitor bacteria evolving in laboratory glassware. For the first time ever (but not without countless units of scientific manpower being expended to this end), the acquisition of a genetically controlled enzyme activity had been recorded in a bacterial population initially totally lacking that activity. As a laboratory vindication of the theory of evolution, the work ranks very high indeed.

Dr Campbell and his group investigated a strain of *E. coli* that possessed all of the machinery necessary to attack lactose except the one specific enzyme that breaks down the sugar molecule. It had the enzyme that admits lactose across the membrane and into the cell, and also the appropriate control systems. But it was defective in lacking the gene, possessed by its parent strain, responsible for producing the crucial lactose-degrading enzyme. Unlike previous investigators, therefore, who had apparently observed the evolution of new metabolic skills that later proved to be merely the release of existing talents that had been repressed, Dr Campbell's group could be sure that if their organism began to ferment lactose, this would be genuine evolution. The appearance of any ability to break down lactose must indicate the emergence of a new enzymic protein — and not the reappearance of an old, latent skill.

Simple, conclusive experiments, prefaced by careful plotting and pondering, so that all ambiguity has been ruled out, are one of the unique joys of scientific research. The work of John Campbell and his friends comes into that category. Having decided on the correct initial conditions, all they had to do was to watch for what now seems to have been inevitable — the evolution of organisms able to break down lactose. They cultivated the *E. coli* in glass plates on agar containing lactose and a little glucose, and an indicator that would reveal any breakdown of the lactose. Routine incubation for two days produced only normal growth. Like all good bacteriologists since Fleming, however, the experimenters kept their old plates, and about a month later there were signs that the lactose was being attacked. One colony showing this activity was transferred on to fresh agar, this time containing lactose as the only source of energy. Again, several new colonies appeared, and one of the faster-growing of these, apparently a good exploiter of lactose, was isolated and transferred once more to fresh lactose medium. The microbiologists repeated the process several times, until there was no further improvement, and turned next to study the new enzyme.

It soon became clear that the enzyme responsible for attacking lactose differed considerably from that in the ancestral bacterium from which the defective strain used in the research had been derived. Its molecular weight and physical characteristics were different. Moreover, by constructing a map of the bacterial chromosome, Dr Campbell and his colleagues found that the new gene was distant from the region in which that associated with the ancestral enzyme would have been. Finally, the new gene did not respond to any of the metabolic control mechanisms that had regulated the old one.

What seems to have happened is that, under strong environmental pressure to attack lactose, as glucose was depleted, a gene producing an enzymic protein for some other purposes (perhaps no longer relevant to the organism) had begun to fabricate one capable of fermenting this particular sugar. Presumably, a copying error during enzyme synthesis had prompted.the assembly of a new type of enzyme with this catalytic activity. The more efficient such an enzyme in degrading lactose, the greater the selective advantage for the bacterium possessing it.

That, of course, is what has long been thought to occur not just among the microbes but throughout organic evolution, as natural selection acts on the variety of novel organisms, with differing abilities and disabilities, that are generated by genetic change. People have even written authoritative textbooks describing the phenomenon in detail. But until 1973, when a strain of *E. coli* in Los Angeles revealed its newfound skill in growing on milk sugar, no one had been able to prove the point.

The origin of complexity

'I used to smile at biologists who spend their entire lives studying the differentiation of slime moulds or the behaviour of flat worms, but I have learned better now,' confessed the molecular biologist and later Nobel laureate, Max Perutz, in 1965. Perutz was writing at a time when the problem of differentiation — the gradual development of the intricate structure of an orchid, an elephant or a sycamore tree from the tiny fragment of DNA in a fertilized seed or egg — was emerging as the greatest challenge to modern biology following the conquest of the double helix. Today, differentiation is at the centre of the stage in biological research. And two groups of microbes have begun to provide some of the answers to the riddle.

This may seem surprising to those who believe that microbes are primitive. What, indeed, does differentiation in microbes amount to? One model system now under intensive scrutiny is the production of spores. The development of ordinary cells of *Bacillus* and related bacteria into the dense and resistant spores that allow them to hold out during bad times — and conversely the germination of spores when the environment becomes more congenial — is itself a complex and wondrous transmutation. Being accessible to investigation in the laboratory, it is also one that offers the prospect of yielding insights into the universal phenomenon of differentiation. Already, bacteriologists have been able to observe sequential changes in enzyme patterns as different materials are synthesized during successive stages in the formation of spores. What they now hope to do is to relate these changes to the underlying genetic control mechanisms that switch various processes on and off according to need.

The second class of microbes helping developmental

biologists with their inquiries are the slime moulds. These are sophisticated cells which aggregate when food is scarce, producing an erect fruiting body that later releases spores that turn into simple moulds once again. The cycle can be manipulated at will in the laboratory, and it too is providing insights into differentiation in 'higher' organisms. One unexpected finding so far is that the signal for the mould to aggregate comes from only one cell in every 300-2,000 or so. They alone can command the population with sufficient authority. Whether this too is a valid analogy with human life is not clear.

Microbes as models in medical research
A decade ago, Dr G. F. Gause, formerly on the staff of the National Institute for Medical Research, London, and now head of one of the institutes of the Academy of Medical Sciences in Moscow, published an important book entitled *Microbial Models of Cancer Cells*. In it he argued that microbial cells are valuable research tools for studying cancer. Microbes do not themselves suffer from this otherwise universal affliction; but they are perfect model systems for investigating the biochemical and genetic processes that accompany the transformation of healthy into malignant cells.

'By investigating microbial models of cancer cells in various micro-organisms it can now be definitely demonstrated how the impaired respiration of the cell results from primary distortions in the nucleic acids,' wrote Dr Gause when the book first appeared in 1965. 'Microbial models also demonstrate distortions in cellular regulatory mechanisms and alterations of the cell surfaces characteristic of many types of tumour cells.' The book was written to provoke interest in the potentialities for cancer research of investigations on micro-organisms. Progress has since been made along that path. The final 'secret' of cancer still eludes us. But whether a single answer exists to the riddle of malignancy, or whether the explanation is more complicated, our eventual understanding of cancer will owe much to studies on microbes and their behaviour.

Another development of potentially enormous value is the use of bacteria to assess chemicals for possible carcinogenicity. A test of this sort has been developed by Dr Bruce Ames of the University of California. Though it is designed basically to reveal substances that can cause mutation, the correspondence between such chemicals and those that cause cancer is so close that it promises to be widely used for screening carcinogens.

Quite rightly, recent years have seen increasing unease, among both the public and research workers, about the use of animals in medical experimentation. Regulations vary from country to country, but everywhere there is growing awareness that, with sufficient incentive, many types of experiments currently conducted with animals could be substituted by alternative methods based on microbes or tissue cultures. The scientists' conventional answer is that, because animals are much more costly to work with than test-tube techniques, they are replaced wherever possible. But against that argument, force of habit and the need to cross-compare results with those of other scientists means that experimental methods using animals are often maintained unnecessarily. If an existing animal technique is available, why should the busy researcher spend time in developing an alternative? In recent years bodies such as the International Association against Painful Experiments on Animals, and FRAME (Fund for the Replacement of Animals in Medical Experiments, set up in Britain in 1970) have been established to stimulate work in that direction. Some of the methods now being investigated exploit microbes whose metabolism resembles that of man or other animals sufficiently to allow measurements relevant to the problem under study. Cancer is just one of the research topics being studied in this way. Test-tube techniques based on microbes have also replaced the injection of pathogenic microbes into animals in the diagnosis of several infectious diseases.

Assays and analyses

Away from the frontiers of research, microbes and their products are extensively employed, day in and day out, by hospital pathology departments and other routine laboratories all over the world. Some of the standard procedures they execute would be impossible without their aid. Others could only be tackled more clumsily, less precisely or at greater cost without microbial assistance.

'Bioassay' is an outstanding example. Chemists use this analytical technique to measure vanishingly small amounts of chemicals. It is the method of choice for many purposes, being more sensitive and simpler than chemical analysis. Bioassay depends upon the capacity of bacteria, yeasts and other micro-organisms to respond to minute quantities of nutrients. Cells of a bacterium unable to grow without biotin can detect as little as 0.000000001 gram of this B-group vitamin. The analyst simply suspends the cells in a tube containing the other nutrients needed for growth, but with the one crucial ingredient missing. He then adds the material to be tested, and incubates the tube. The amount of growth after a standard interval is an index of the concentration of biotin in the test material. Growth is measured by passing light through the tube and into a photoelectric cell, which registers the strength of the emerging beam and thus the turbidity of the liquid. The technique is widely used in assaying vitamins, amino acids and other nutrients. Similar methods are employed to assess the potency of antibiotics.

Microbial enzymes too have many routine applications. One is to investigate the structure of molecules. As in steroid transformation, the ability of specialist enzymes to effect small changes at particular points on a molecule can provide information about the way in which the various atoms in a molecule join together. In pathology laboratories, microbial enzymes are used regularly in measuring quantities of normal and abnormal substances in blood and other body fluids. With the introduction of immobilized enzymes (p. 162), these applications will certainly increase. A simple, cheap and extraordinarily efficient gadget was introduced recently in the US which employs a bacterial enzyme to measure glucose in blood and urine. Its chief value is in diagnosing and monitoring diabetes.

All of these techniques, glamorous and routine, underline the unity of life. If micro-organisms and men differed substantially in their biochemistry and genetics, it would be nonsense to think of using microbes to investigate human cancer, human development or human diabetes. But the likeness is inescapable. Even our apparently most contrasting functions, when examined closely, prove to be similar. And far more significant than such distinctions are our profound and fundamental areas of identity.

10 Infection in Perspective

One might expect Dr René Dubos — distinguished bacteriologist, public health specialist and former pupil of Selman Waksman — to believe in unmitigated microbial malevolence. Having spent most of his life seeking methods of dealing with infectious illness, he could well have concluded (as have many medical microbiologists) that the microbial world is totally antipathetic to human life and that its most typical members are those that cause typhus, cholera, tuberculosis and plague.

Not so. In his book *The Mirage of Health*, published in 1960, Dubos writes:

> The knowledge that micro-organisms can be helpful to man has never had much popular appeal, for men as a rule are more preoccupied with the dangers that threaten their life than interested in the biological forces on which they depend for a constructive existence. . . . Plague, cholera, and yellow jack have found their way into the novel, the stage, and the screen, but no one has made a success story out of the useful role played by microbes in the intestines or the stomach. . . . The world is obsessed — naturally so — by the fact that poliomyelitis can kill and maim several thousand unfortunate victims every year. But more extraordinary, even though less dramatic, is the fact that millions upon millions of young people become infected with polio viruses all over the world, yet suffer no harm.

Infectious disease is real enough. Though the overwhelming thrust of microbial endeavour on earth is constructive, and directed towards our health and wellbeing, there is little point in denying either the scourges of plague and typhus or our occasional visitations by influenza and the common cold. But if we abandon our anthropocentric

viewpoint for a moment and examine infectious disease
against a wider biological and social perspective, it appears
in an altogether different light.

What causes tuberculosis and cholera?
According to orthodox theory, taught in medical schools
and printed in textbooks, tuberculosis — the 'white
plague' — is a wasting disease, often affecting the lungs,
whose origin was unknown until 1882 when the great
microbe-hunter Robert Koch proved that a bacillus,
Mycobacterium tuberculosis, was responsible. Koch laid
down four strict conditions (later known as Koch's postu-
lates), each of which must be fulfilled before a micro-
organism can be conclusively incriminated as the agent
of a particular disease. The microbe must be invariably
associated with all cases of the disease. It must be isolated
as a pure culture, and allowed to multiply over repeated
generations. When inoculated into a susceptible animal,
the organism must again cause the disease. Then it must
be isolated in pure culture once more from the diseased
animal. Koch fully satisfied these criteria in his own work
on tuberculosis. The meticulous paper in which he reported his studies amounted to as exhaustive a case against
the tubercle bacillus as one could possibly imagine.

Yet Koch's interpretation was far from fully accepted at
first. The obvious association of tuberculosis with poverty
and malnutrition provided a much more satisfying explanation of the aetiology of the disease. Many folk, doctors
included, continued to believe that these two factors had
far greater significance than any invisible microbe. Besides, did not improvements in living conditions generally
lead to a reduction in the incidence of consumption?
Certainly, tuberculosis was disappearing long before the
microbe-hunters came along. And did not bracing air
and nourishing food help to repel the disease?

Early antagonism towards Koch's interpretation is
usually written off, in books on the history of medicine,
as tiresome obscurantism in the face of overwhelming

scientific evidence. On the contrary, it was altogether reasonable. For in a population where the tubercle bacillus is ubiquitous, everyone becomes infected, but few are troubled. There is a high level of natural resistance in the community and only people who are debilitated through malnutrition, or perhaps exhaustion owing to over-work, develop tuberculosis. Most of those infected shake off the bacillus easily enough and acquire even greater immunity. In a real sense, therefore, malnutrition and emaciation are the actual cause of tuberculosis — in that context. The situation is very different in a community that is uniformly well-nourished and sanitized and where the tubercle bacillus has been driven to virtual extinction. Here natural immunity will have waned to the point where no one has any resistance against tuberculosis, while variations in nutrition are a less significant determinant of disease. A virulent strain of *M. tuberculosis* might well attack a sizeable proportion of people, causing full-blown consumption. But that has not been the situation for most of recorded history. At most places and times tuberculosis has been 'a social disease with medical aspects', as William Osler called it. Osler was a firm believer in the germ theory, but he realized that it portrayed a grossly distorted view of infectious disease.

The same is true of cholera. One of the most famous episodes in the history of preventive medicine occurred in 1854 when John Snow, Britain's first professional anaesthetist, demanded that the handle be removed from a pump in Broad Street in Soho. A cholera epidemic was raging in London and had caused as many as five hundred deaths in ten days in one small area alone. After making house-to-house enquiries, Snow discovered that the victims had all collected their drinking water from the same Broad Street pump. The handle was duly disconnected and a few days later the epidemic abated. No more new cases developed. But the medical profession did not at once accept the implications of Snow's action. Many of its members remained unmoved even by Koch's discovery

of the cholera bacillus in 1883. They were more impressed
by the fact that cholera, like tuberculosis, was invariably
associated with filthy living conditions.

One of the most intriguing characters in the evolution
of medical science was a Bavarian doctor, Max von
Pettenkofer. Though considered to be the founder of
modern ideas of hygiene, he never accepted the germ
theory of disease. The theory became firmly established
between 1860 and 1890, but von Pettenkofer resisted until
the last and died in 1901 still convinced that 'miasmata'
rather than microbes were the cause of many diseases.
He believed in putting his ideas to the test and in 1892,
when aged 74, he drank several tumblerfuls of cultures
isolated from fatal cases of cholera. At about the same
time the Russian pathologist Elie Metchnikoff did likewise.
So did several of their colleagues. Some of these intrepid
experimenters experienced mild diarrhoea. All had enormous numbers of cholera bacilli in their faeces. But none
developed anything like cholera.

Clearly, the story of infectious disease is more complex
than the early microbe-hunters suspected. The simplistic
version of their work, which still forms the basis for many
popular accounts of infectious disease, is that they identified the specific causes of a succession of diseases, and
thus initiated further research to devise specific remedies.
But this outlook led some of the crusaders and their
followers to minimize the influence of nutritional and
environmental factors on infectious illness. It placed far
too much blame on microbes as agents of disease, and
has ever since engendered a grossly reductionist picture
of man and microbes locked in unremitting conflict.

Strengths and weaknesses of specific aetiology

The gospel propounded by the microbe-hunters was
termed the 'doctrine of specific aetiology'. It stated simply
that specific microbes caused distinct diseases. This is
an unremarkable concept today, but it was immensely
powerful and productive when first enunciated. Merely

to begin to distinguish clearly between different infections that had been confused for centuries constituted a great step forward. When this was followed by the isolation of a rogues' gallery of infamous microbes — the tubercle bacillus in 1882, the cholera bacillus and the streptococcus of erysipelas fever in 1883, the diphtheria and tetanus bacilli in 1884 and the pneumococcus in 1886 — it was tempting to overlook the possibility that the theory had any limitations. And when, in 1911, Paul Ehrlich introduced a specific remedy (Salvarsan) for syphilis, the now fully vindicated theory became a dogma.

Ironically, it was in other areas of medicine that specific aetiology proved most trustworthy. The specificity revealed initially by the microbes was well in tune with the deterministic mood of nineteenth-century science (and with the economic determinism of Marx and Engels). After early claims for its successes in explaining infective conditions, hopes grew that a diligent search would reveal the specific causes of all diseases, not just infectious ones. Two fields in particular yielded to this approach. The first was vitamin deficiency diseases, where all the varied symptoms and effects of conditions such as scurvy and rickets could be ameliorated by provision of a single chemical compound. The relationship of insulin to diabetes is another case in point. The second dramatic vindication of the theory came from inherited metabolic disorders, which were discovered by Archibald Garrod at the turn of the century. These are diseases such as phenylketonuria, caused by specific genetic defects that prevent their corresponding enzymes from operating.

Interactions between man and microbe are much more complex, less deterministic. Von Pettenkofer's good health after swallowing cholera bacilli proves as much. So does the fact that many of those in the room when Robert Koch read his historic paper on tuberculosis in 1882, like most adults at the time, must have been infected with the tubercle bacillus, without harm, earlier in their lives. Tests that Koch carried out on himself eight years later proved that he had been infected at some time with *M. tuberculosis.* But he never suffered from consumption and remained in vigorous health until he died of a stroke at the age of sixty-six.

The main reason why the theory of specific aetiology was cast into such a rigid form was the artificiality of the laboratory models used by the pioneers. Early successes of experiments in which they injected animals with large doses of pure cultures of micro-organisms encouraged the belief that microbes are both the necessary and sufficient causes of certain diseases. This led to unnecessary controversy between the proponents of the new scientific theory and those who were more impressed by the influence of environment, nutrition and similar non-specific factors. There are, indeed, few if any diseases for which the dogma of specificity has provided anything like a complete explanation. While paying tribute to its successes, René Dubos points out that the dogma has failed to account for any of the major medical problems of our times — cancer, cardiovascular disease and mental illness. 'In reality, the search for *the* cause may be a hopeless pursuit because most disease states are the indirect outcome of a constellation of circumstances rather than the direct result of single determinant factors.'

An ecological view of infection
One of the most terrible pestilences the world has ever known was the potato blight that ruined the economy of Ireland in the 1840s. The fungus responsible, *Phytophthora infestans*, caused appalling famine and economic disaster. In ten years, Ireland's population fell by three million people — a million dead and two million emigrants to America, Canada and other parts of the British Empire. Yet today the same fungus is still prevalent in the potato fields of Ireland and the rest of the world. We have learned

to live with *Phytophthora*, by using farming methods
that allow both it and the potato to live in mutual
harmony.

The natural history of myxomatosis shows that it is
not in the long-term interests of any microbe to ravage
another species of life with lethal ferocity. Strains of
myxomatosis virus liberated among unwanted populations of rabbits did just that initially, but soon the virus
began to evolve towards a much milder form, in which
it no longer destroyed its hosts. Most of the great plagues
and pestilences of history, in fact, can be traced to accidents
or human interference that have disturbed the balance
of natural populations. When we adopt a broad, ecological view of the natural history of what we now categorize
as infectious disease, the simplistic notions of specific
aetiology seem severely blinkered. What we see instead
is a world in which coexistence between man and microbe
is not only possible but also more natural than the battle-
ground view portrayed in so many accounts of infectious
disease. When this balance is disturbed, to our disadvantage, we usually have only ourselves to blame.

Sir Macfarlane Burnet, the Australian immunologist,
points out that many and possibly all of our most lethal
infectious diseases are naturally infections of other verte-
brates or insects, which we have stumbled upon by accident. Plague, rabies and yellow fever all come into this
category. Typhus fever is the classic example. Originally
a mild infection of rats, man encountered typhus only
through accidents of history.

War epidemics of typhus have always been associated
with famine, hardship and dislocation of social life
[Burnet observes]. Under happier, if still crowded and
filthy circumstances, typhus can take on a milder
character. Following the general rule amongst infections, typhus is far milder in children than amongst
adults, and in endemic regions such as the cities of
North Africa a high degree of immunity may be devel-

oped by such mild childhood infections. Here we have reached the end of the process of transfer from rat to man. The mild disease of the rat passes accidentally to man, it finds a new vector in the louse, and under circumstances of war and famine spreads widely and fatally, but where circumstances allow its easy spread in a stationary population, it eventually develops the character of a typical relatively mild endemic infection.

Even Hans Zinsser, in his portrayal of infectious disease as 'a nationalism of species against species', suggests that typhus fever in Europe originated during the long-drawn-out wars of the sixteenth century in Hungary when Christians and Moslems alternately conquered and reconquered the country. There is no evidence of typhus before that time, but every war in Europe since has led to serious epidemics. One of the most appalling epidemics ever known was the typhus that swept through Russia during and after the 1917 revolution.

Polio and cleanliness
The greater severity of first infection in adults than in children is vividly illustrated by the story of poliomyelitis. This is the outstanding but by no means only situation in which superior hygiene and sanitation, by postponing infection from infancy to adulthood, have transformed a relatively unimportant infection into a devastating disease. Contrary to the dogma of specific aetiology, poliomyelitis did not decline as Western civilization adopted increasingly scrupulous standards of cleanliness. Instead it became more prevalent and more serious. Statistics unambiguously confirm the grim picture. In 'natural' surroundings, all children encounter the virus in the very earliest years of life. It enters the body by mouth and multiplies in the throat or lining of the intestine. The body responds by producing antibodies and becoming immune. The vast majority of infected children experience no ill effects whatever. At most they have a

transient sore throat. In rare instances, however, the virus passes into the blood and thence into the nervous system. This may cause nothing more than a headache, but a few less fortunate victims develop 'infantile paralysis'.

By ensuring that most children escape early infection, modern hygiene turned this picture of universal infection and rare disease into one of perilous, epidemic proportions. Serious polio outbreaks have been directly related to improvements in living standards. The first known epidemic was in Sweden in 1887, and for forty years or so they were confined almost entirely to Scandinavia, the north-eastern United States, Canada, Australia and New Zealand. By the 1950s outbreaks were more widespread. And in the late 1960s, when the prevalence of polio had fallen dramatically in the affluent West because of immunization, the number of cases was rising rapidly in many developing countries, in parallel with improvement in living conditions. The overall pattern of all these epidemics has been for older individuals to be affected in successive outbreaks.

Even comparatively recently, the extent of 'invisible' infection, causing no illness or inconvenience whatever, has been remarkable. During the early 1950s, virologists in several countries detected the polio virus, circulating throughout the community in late summer whether or not an epidemic was in progress. There was great alarm at that time because research workers could find the virus without difficulty in sewage from any of the main outfall channels of cities such as New York and Chicago. Even when a few paralytic cases did occur, they were grossly overshadowed numerically by widespread symptom-free infection. When epidemiologists turned their attention to Cairo, Bombay and the native 'locations' of Johannesburg — places which combined a warm climate with low standards of infant care — the propensity of the virus to disseminate widely through the infant population while causing only rare disease became even more obvious.

Infection is not disease

The conclusion from our experience with polio (and from Dr von Pettenkofer's bizarre experiment on himself) is that there is all the difference in the world between infection and disease. We are inclined to believe they are the same. That is why we go to inordinate lengths to avoid 'infection' — and why, conversely, we are comforted to know that we have a genuine illness when a doctor pronounces (almost always without any evidence whatever) the magic formula: 'virus infection'. But the truth is that we are all infected continually — and not just by do-gooders like the gut bacteria that synthesize vitamins. Most of us harbour in our tissues, either intermittently or for our entire lifetimes, many microbes that have the capacity to paralyse, poison, starve, suffocate or bleed us to death. People who are in every sense hale and hearty can carry tubercle bacilli, polio viruses, the streptococci that cause rheumatic fever and many other 'pathogens' without being inconvenienced in the slightest way.

Several factors account for this surprising state of affairs. We may possess antibodies that hold microbes in check. Cold sores, produced by herpes simplex virus, are a perfect example of symbiosis in which the virus lives contentedly in the presence of antibodies. Herpes simplex could induce a severe and potentially lethal infection, but it doesn't. The antibodies could destroy the virus, but they don't. Instead, herpes simplex and its host co-exist quietly, the virus making itself manifest only occasionally as a result of its partner's sunburn or raised temperature.

Probably more significant in accounting for co-existence is the 'herd immunity' that we develop as a species when exposed continually to a particular microbe. By mutual adaptation, man and microbe have evolved arrangements that are sensible for the survival of both. Many historical examples bear this out. During the eighteenth and nineteenth centuries, epidemics of smallpox, tuberculosis and even measles, brought by European invaders and explorers, cruelly decimated the Amerindian and Polynesian populations. Yet within a few generations the mortality associated with these diseases began to fall spontaneously — without any medical or other measures being taken. Similarly, African trypanosomiasis, when first introduced into new regions of Africa, killed up to two-thirds of the population. The disease still persists, but it now causes few deaths.

Nutrition, general fitness and probably also that mysterious complex of what we label 'psychosomatic' influences also bear upon the outcome of any encounter between micro-organisms and man. We are undoubtedly more resistant to infectious disease when we are fit, well nourished and contented than when hungry, exhausted and under emotional stress. The importance of nutrition is exemplified by the steep increase in skin and other infections among prisoners-of-war kept under bad conditions in Germany during the last war. Many of the microbes responsible were undoubtedly present on the prisoners' bodies before confinement, when the men were receiving adequate food.

Though the dogma of specific aetiology would exclude such ideas totally, it also seems likely that psychological factors can trigger off the conversion of infection into disease. Even in animals, 'travel fever', owing to vague infectious illness that comes to light merely as a result of moving livestock from one environment to another, is a well recognized phenomenon. Psychosomatic medicine has, however, only recently graduated from being a nutty fringe interest to become a serious research topic. In time, it may well shed light on the presently obscure but well founded link between infectious disease and mental wellbeing.

Welcome residents

The beneficence of infection is becoming clear from studies on microbial inhabitants like the intestinal bacteria and the scavengers of the anus. The most specific example known is that of the vagina. From after the first month of life until puberty, a girl's vaginal secretion is slightly alkaline and harbours only a sparse population of microbes. During adolescence, however, glycogen (animal starch) appears in the wall of the vagina, as a result of hormone activity. This alters the microbial flora radically. The vagina now becomes colonized by several bacteria, principally 'Döderlein's bacillus', and by yeasts. Doderlein's bacillus resembles the bacteria that turn milk sour by producing lactic acid. It begins to ferment the glycogen, making the vaginal secretion acidic, which is important in preventing infection by undesirable microbes. After the menopause, glycogen disappears, the microbial flora returns to what it was before puberty, and the secretion becomes alkaline once more.

The mechanism is a protective one, evolved to shield the vagina and uterus from infectious disease during the period of sexual activity and child-bearing. It is ultimately under the control of the female sex hormones, and thus perfectly geared to operate at precisely the time in life when the microbial population of the female genital tract is most critical. But Döderlein's bacillus is far from being a ruthless scavenger that obliterates all other signs of microbial life. What it does is to encourage and sustain a a mixed, balance population of microbes admirably suited to the task in hand. This natural population is gravely threatened by ladies who foolishly spray, douche, dowse or dust themselves with vaginal deodorants and anti-septics. The most likely effect of such misguided measures is to destroy the vagina's microbial defence machinery and thus to allow invasion by totally unwelcome organisms.

Co-operation out of conflict

In 1972 a Japanese biologist working at the University of Tennessee reported experiments in the laboratory that confirm what has also become clear from a study of infec-tion in nature — that microbes favour co-existence, not conflict. Like Dr John Campbell's work, which revealed evolution in a test-tube (p. 187), Dr Kwang Jeon's research portrayed in vignette the emergence of symbiosis among microbes that were formerly enemies. The two microbes investigated were a bacillus and an amoeba. At first, the bacterium invaded the amoeba, prevented it from growing and indeed almost destroyed it. But after living together for some years in a situation of uneasy compromise, the bacterium became not only harmless to the amoeba but positively beneficial. By transferring nuclei from infected colonies into non-infected colonies, and vice-versa, Dr Jeon was able to show that the bacterium had become inoffensive and that the nuclei of the amoebae were now dependent on the bacterium if they were to function normally. The experiments are the first to have demonstrated the conversion of a parasite into a symbiont, living in perfect harmony with its partner, within an observable period of time.

Germs are good for us
The dependence of animals on the cohorts of microbes that normally inhabit their bodies has been amply confirmed recently by work in the developing discipline of 'gnotobiology'. Methods of raising pigs, rabbits, mice and other animals entirely free from microbes were brought to perfection in the early 1960s. Research on these pristine animals, and on the behaviour of germ-free animals artificially inoculated with single, defined strains of microbes, is termed gnotobiology. This new science has already demolished any doubts that may have persisted about the importance of the teeming microbial life found in and on man and other animals — particularly that in the intestine. As Dr Theodor Rosebury puts it: 'The germ-free animal is, by and large, a miserable creature, seeming at nearly every point to require an artificial substitute for the germs he lacks.'

One defect of animals reared without microbes is scarcely surprising: they are extraordinarily vulnerable to infection. The defensive lymphatic system is poorly developed, and there are gravely inadequate quantities of immunoglobulins, the proteins of which antibodies are made. Normal rats, for example, become resistant to anthrax very early in life: whereas at birth the number of anthrax spores in a dose necessary to kill half of the animals in any one experiment is 10,000 (this is the conventional way of measuring the lethality of germs and poisons), very soon the figure rises to 1,000,000,000. Germ-free rats do not acquire this resistance, however, after birth or at any time. Similarly, if germ-free guinea pigs consume water containing dysentery bacilli, they succumb rapidly and die: ordinary animals are totally protected against the disease. In neither of these cases do pathogenic bacteria in the environment induce the resistance by stimulating production of specific anthrax or dysentery antibodies. The animals' normal, 'non-pathogenic' flora is responsible.

The probable explanation is that the microbial population, well adjusted to life in the intestinal tract, prevents the unwanted bacteria from gaining a foothold. Without that flora, pathogens can establish themselves quickly and disastrously. Even using massive quantities of cholera bacilli, in the style of von Pettenkofer, it is extremely difficult to infect ordinary, microbe-ridden animals with cholera. But the same organism readily invades the gut of germ-free animals. Many bacteria that are usually totally harmless also colonize pristine intestines and produce disease.

Less predictable characteristics of germ-free animals are defects in the structure of their intestinal tract, and their need for nutrients that animals normally receive from their resident microbes. The gut abnormality resembles those seen in human diseases like sprue, in which the gut no longer absorbs food materials properly. The wall of the small intestine is thin, with little connective tissue. The villi (tiny finger-like projections through which sugars and amino acids pass into the bloodstream) have a bizarre shape, and their cells are replaced less rapidly than usual. Clearly, bacteria are essential for the normal development of the intestine. Otherwise it becomes weedy, inefficient and unresponsive to mechanical stimuli.

The penalties of pristine sterility are even more severe for rodents, which usually harbour a rich contingent of bacteria and protozoa in the caecum. Germ-free guinea pigs, rats and rabbits have a grotesquely enlarged caecum that cannot function properly. Lacking the stimulus to evacuate, normally caused by its microbes, the caecum fills up continuously. Its muscular contractions, which churn and push the food along, are feeble and sluggish. But all of these irregularities are potentially reversible. Feeding normal intestinal bacteria to germ-free guinea pigs restores the bowel's healthy behaviour and architecture within a few weeks.

Without their microbial cohabitants, germ-free animals also have abnormal nutritional needs. As well as manufacturing vitamin K, vitamin B12 and probably other vitamins of the B group, bacteria living in the gut share several other synthetic responsibilities. They help to release 'waste' nitrogen, in the form of ammonia, from amino acids. They modify bile acids and other steroids, and play a part in the metabolism of polyunsaturated fatty acids. Bacteria in young piglets produce the enzyme needed to digest lactose in the mother pig's milk; germ-free piglets need glucose in the diet or else succumb to hypoglycaemia. Another recent discovery is that bacteria in our intestines break down food additives such as colouring and flavouring agents. The soil is not the only location where microbes operate as scavengers, coping with our chemical debris!

Microbial aid of this sort has come to light only over the last few years, and microbiologists now realize that a considerable amount of further research is required to clarify the underlying mechanisms. Particular organisms responsible for many of the beneficial effects have yet to identified. This is a tricky task because several of the normal gut inhabitants are strictly anaerobic, and thus difficult to culture in the laboratory. The real challenge, however, is to study the ecology of the normal, richly varied population of the gut. The very opposite approach — gnotobiology — has yielded many surprising discoveries that are relevant to that study. We now know, for example, that even the activity of the heart (as measured by the volume of blood being pumped) is weaker in germ-free animals than in those with their healthy quota of micro-organisms. Only gnotobiology could have revealed this totally unexpected influence on cardiovascular development. But only studies of the normal, mixed microbial flora — as urged by Serge Winogradsky — will reveal the full strength and detail of its contribution to good health.

How not to use antibiotics

The earliest clues that our resident microbes might be important came from the days when antibiotics were first introduced and were grossly misused, causing repercussions, ranging from vitamin deficiencies to vaginal thrush and yeast infections of the intestine. These conditions occurred because doctors were wielding antibiotics with lavish abandon, knocking out whole populations of bacteria and thus disturbing beneficial functions and the balanced relationship of man with his microbes. Friendly, neutral and hostile germs were slaughtered indiscriminately.

Two or three decades later one might have expected the lessons of that period to have been thoroughly learned. Not so. Doctors still prescribe antibiotics unnecessarily for trivial as well as serious conditions. Potent drugs vital to the treatment of occasional serious infectious illness are freely available across the counter in many countries. And farmers are allowed to put some of the same antibiotics in their animal feedstuffs to promote growth. The inevitable result of all of this has been a menacing and worldwide proliferation of strains of bacteria, both pathogenic and non-pathogenic, that are resistant to antibiotics.

If ever there was a topic that needs the insights of ecological wisdom, this is it. Antibiotics are natural microbial products which have an important and legitimate use in combating severe infectious disease. Employed selectively, they have richly justified their original reception and reputation as wonder drugs. But when deployed as crudely and sloppily as the chemical industry's pesticides and fertilizers have often been in the past, the consequences have proved to be horrendous. In recent years, epidemics of untreatable disease, produced by bacteria insensitive to the antibiotics that would otherwise have been used in therapy have become increasingly common. In Guatemala alone, in 1974, twelve thousand people died from dysentery caused by a resistant bacillus. This is not a natural situation, a result of microbial malevolence — it is a direct outcome of our own crudity in using microbial products.

Antibiotic resistance arises in bacteria in two different ways. The first is classical Darwinism. Even in a population of bacteria not previously exposed to a particular antibiotic, occasional organisms able to withstand its action will appear. They emerge by spontaneous mutation. Through a process of 'survival of the fittest' — accentuated by rapid multiplication — such bacteria will tend to become dominant when the antibiotic concerned enters their environment. Much can be done, however, to reduce the risk of this happening — principally by using the right quantity of drug, at the right time, to deal a death blow to an infection as quickly as possible. Conversely, indiscriminate application of antibiotics — especially giving them in inadequate doses and/or over long periods of time, and for trivial infections where they are not genuinely needed — will promote the emergence of resistance.

The other type of drug resistance is transferable between bacteria and was first discovered by Japanese bacteriologists in 1959. During brief contact, a resistant bacterium can confer resistance on a previously vulnerable organism by donating a plasmid. In this way, a whole population of organisms initially susceptible to, say, penicillin can quickly become protected. Moreover, resistance is often multiple: sensitive bacteria may become invulnerable to as many as seven or eight different drugs. Transferable resistance is particularly common among bacteria that colonize the gut of man and animals, including not only agents of disease, such as dysentery and typhoid fever bacilli, but also *E. coli*. Resistance can be transmitted between pathogens and non-pathogens, and this raises the possibility that harmless microbes will provide a 'pool' of resistance that may at any time be passed on to virulent ones.

Transferable resistance is a natural phenomenon. We have turned it into a serious problem. The use of antibiotics in factory farming has provided virtually ideal conditions for resistance to flourish and spread. One study, conducted in Holland from 1959 to 1969, showed a striking association between the deployment of particular antibiotics in farming and the increasing prevalence of correspondingly invulnerable strains of *Salmonella typhimurium*, which can cause food poisoning in both man and farm animals. In Britain Professor E. S. Anderson, Director of the Enteric Reference Laboratory at Colindale, London, has reported similar findings. They include direct evidence of a huge rise in drug resistance stemming from the massive application of ampicillin and other drugs in the intensive rearing of calves. When farmers deploy antibiotics in this way they are not giving them, legitimately, to treat disease: they are using them to boost the growth of their livestock (how the drugs do so is still a mystery) and to prevent the onset and spread of disease — something that can be achieved much more soundly by improving the conditions of husbandry.

Some limitations have been placed on antibiotic usage in agriculture. Confronted by disquieting evidence of public health hazards, Britain in 1969 and the US in 1972 decided to restrict the freedom of farmers to include certain drugs routinely in feedstuffs. But this is not a problem that can be tackled completely on a national basis.

Microbes do not recognize political frontiers, and one country's prudence with antibiotics can be made meaningless by the profligacy of its neighbours. A major problem is financial. Broiler fowls fattened by antibiotics are slightly cheaper than they would be otherwise. The fact that the potential health risks totally eclipse such minor benefits has as yet been overlooked by the authorities in many countries. Even the World Health Organization — arguably the UN's most successful agency — has failed to do more than monitor and lament the worldwide proliferation of drug-resistant bacteria.

Farmers apart, both doctors and patients bear some of the blame for the current situation. Many of us expect, and many doctors provide, antibiotics when they are quite unnecessary. The vast majority of sore throats, tummy upsets and similar trivia clear up without any treatment whatever. It is actually better that they should be dealt with by the body's natural defence mechanisms, which are thereby strengthened, rather than be doused in drugs. Besides, many everyday gastric and respiratory troubles are caused by viruses, which are unaffected by antibiotics anyway. Even *Salmonella* food poisoning is, for most of us, a mere nuisance, which does not require chemotherapy. But in the very old, the very young and people debilitated through other causes it can become seriously complicated and even prove fatal. In such patients, antibiotic treatment is essential. It is in those cases that the failure of micro-organisms to respond to the whole range of antibiotics is the most tragic.

No need to exaggerate
Even with typhoid fever, a serious and potentially lethal infectious disease, we need a sense of proportion. Typhoid is nothing like the mysterious, utterly devastating plague that is suggested by the inevitable newspaper headlines whenever there is even a suspected case, let alone a genuine outbreak. 'Typhoid children banned from school for ever' ran a caption in the *Daily Telegraph* of 17 November 1971.

When typhoid appeared in the Swiss ski resort of Zermatt in 1963, the panic and hysteria generated were greater still, and certainly out of all proportion to the severity of the disease. One Englishman who had been in Zermatt during the outbreak, but who had not been infected, returned home to find himself and his family ostracized. He was excluded from work, his wife had difficulty in buying food from a local store and people objected to his children attending school.

It was in response to this type of nonsense that Professor E. S. Anderson — who, as head of the national laboratory in Britain devoted to monitoring typhoid and related bacilli, has greater reason than anyone to welcome public concern about the disease — felt obliged to publish an article entitled 'The truth about typhoid'. In the article, which appeared in *The Times* of 31 December 1971, he pointed out a number of basic but forgotten truths about the disease. First, the incidence of typhoid fever in a country like Britain, with good hygiene and water supplies, is very low, as are the chances of it spreading. The bacillus cannot be transmitted by contact or by coughing or sneezing but only by contaminated food or water. The illusion of epidemic spread is, however, easily created when people infected at the same time have differing incubation periods and thus fall ill after different intervals.

Secondly, typhoid fever is now easy to treat, using the antibiotic chloramphenicol. Even in the famous Aberdeen outbreak of 1964, which attracted unique and phrenetic coverage by the news media, there were only 3 deaths out of 507 cases, and at least one of those deaths was of an elderly woman who was already gravely ill before she contracted typhoid. The importance of chloramphenicol as treatment for typhoid fever is one outstanding reason for taking action to prevent indiscriminate use of this drug, leading to the emergence of resistant strains of the typhoid bacillus.

Another needless fear concerns typhoid carriers. Only a tiny percentage of people who contract the disease become carriers, and only very rarely do they later transmit the disease to other people. The few carriers who have been found to do so have usually been elderly ladies who acquired typhoid fever many years earlier and who have become sources of infection when their standards of personal hygiene deteriorated in old age. One lady, who caused a minor outbreak in 1948, had been infected in 1895. She had worked as a schoolmistress all her life but only once, in 1926, had she apparently infected two visitors. As Professor Anderson pointed out, even she could hardly be described as an active and serious threat to the community. When carriers are detected, they are instructed in special hygiene precautions, and while they are not allowed to work in restaurants, the food industry or water works, they can live perfectly normal lives without posing any danger to others.

As to the risk of typhoid as a result of travel to places like the Mediterranean basin, where the disease is not uncommon, the hazard — which is small anyway — can be reduced to a negligible level by TAB inoculation every three or four years, by avoiding foods such as green salads and raw shellfish, and by avoiding unsterilized water. The chance of typhoid fever spreading from one patient to the general population is also negligible. Typhoid epidemics are easily controlled and usually self-limiting.

In sum, Professor Anderson concludes, typhoid fever has been allowed to generate far too much alarm and consternation: [There is]

> no need for panic about typhoid; no need to ostracise contacts; no need to fuss about carriers, most of whom only very exceptionally transmit the infection and who, once detected, are easily controlled; and least of all is there any justification for the sensationalism to which this relatively rare and quite tractable disease is submitted by the media of mass communication.

Needless antipathy
But the news media are not the only culprits, Doctors too are still unduly impressed by the nastiness of microbes, and thus feel they must do something — anything — to

assist their patients in repelling the dangerous beasts. Deploying antibiotics with reckless abandon is one example of this behaviour. Another, and even more curious, is the way in which doctors (contrary to a reliable medical axiom stating that one should deal with diseases rather than merely their symptoms) prescribe drugs such as aspirin to inhibit fever during influenza and other virus infections. In doing so, presumably, they believe they are helping to fight the infection. In practice, they may well be preventing the body from dealing with viruses in its own way.

André Lwoff, Nobel laureate and one of France's most experienced virologists, has been arguing against this routine medical practice for many years, though with little success. Medical practitioners see fever as an intrinsic part of disease. Virologists like Professor Lwoff know that raised temperature is one of the most effective means of limiting virus invasion in both animals and plants. Lwoff's own studies have shown that polio virus grows well in tissue culture cells at 95 degrees fahrenheit but increasingly poorly as the temperature approaches 104 degrees fahrenheit. Between 101.3 and 102.2 degrees fahrenheit the release of new infectious virus particles falls from 95 per cent to 2 per cent of that at 95 degrees fahrenheit. Thus even a very slight difference in body temperature can determine the course of a virus infection. Professor Lwoff and his colleagues at the Pasteur Institute in Paris also found that when they gave a drug to lower the fever of dogs infected with vaccinia (the virus of smallpox vaccine) the mortality markedly increased. In another series of tests, mice suffering from virus encephalitis could be rescued by a massive injection of antibodies, but only if injected on the first day. Raising the animals' temperature was equally effective, even on the second or third day.

Research reported in 1969 by Dr C. Scholtissek and Dr R. Rott, of the Institut für Virologie, Justus Liebig-Universität, Giessen, provided a possible explanation for these effects. They found that a slightly high temperature inhibits the assembly of at least one type of influenza virus. A crucial enzyme involved in manufacuring new virus particles in infected cells is unstable at the sort of temperatures reached during an attack of the disease itself. Moderate fever may well be one of the most important natural defence mechanisms against influenza, and one that doctors should not try to thwart. Of course, substantially raised temperature can be a serious matter, but the indiscriminate use of aspirin to combat every trivial febrile illness is quite unjustified. It almost certainly does more harm than good.

Envoi

Most microbes are helpful, not harmful. Most infections do not cause disease. Most infectious diseases do not need to be treated. Believing otherwise, and acting on those misconceptions, we have made it more difficult to deal with serious infectious disease, even when using antibiotics, those powerful weapons that microbes have provided.

11 Extinction or Exploitation

One of the most off-beat new ideas dreamed up recently for using microbial talent is the monitoring of air pollution by lichens. Several species found on tree trunks, walls and gravestones are exquisitely sensitive to traces of heavy metals in the atmosphere. Two others, *Physcia caesia* and *Xanthoria parietina*, cannot live in areas with a mean annual sulphur dioxide concentration above 0.05 parts per million. By mapping their occurrence, therefore, it has been possible not only to plot regions of high and low pollution, but also to assess the influence of topography and wind direction on the dispersal of invisible effluvia. By recording the luxuriousness or feebleness of growth of several species, teams of observers have charted atmospheric pollution over the whole of Britain. The technique, while soundly dependable, is absurdly cheap. And it requires neither skilled manpower nor sophisticated laboratory equipment. Students of Britain's Open University, living at home throughout the country, have taken part in a national air pollution study based on observations of lichens; so too have fifteen thousand schoolchildren, in an exercise organized by the Advisory Centre for Education.

It's a nice story. Yet it has a darker aspect. The very sensitivity that makes lichens so valuable in detecting and quantifying our industrial misdeeds also means that these precious microbial associations are gradually disappearing — and not only in areas of gross environmental spoilage. Omnipresent pollution, together with our habit of destroying forests and mature trees wherever they are to be found, are combining to drive many lichens inexorably towards extinction. Few people care, and it was to focus attention on the appalling situation that Dr David Richardson wrote a book (published towards the end of

218

1974) called *The Vanishing Lichens*. Though he ended the book with a hope that such a title would soon no longer be appropriate, he warned that this would require a complete reversal of the present destructive trend, together with widespread recognition of the reasons why we should try to conserve the lichens.

Man's relationship with the lichens is instructive. Contrasting with our total apathy about their impending extinction, even casual observation reveals a staggering range of different roles that these crusty but gentle microbes play, have played and could play in human affairs and natural ecology. First, they are an important source of food. They form the major constituent of diet for reindeer in Eurasia and for caribou in North America during the first six months of winter when little or no green vegetation is available. Other deer, including musk-oxen in the high Arctic, eat smaller quantities. Rather than seeking out lichens with the greatest protein content, deer prefer those with a sizeable proportion of complex carbohydrates, which are broken down to sugars by enzymes in the stomach, aided by rumen bacteria. Mites, insects, land-snails and slugs also feed off lichens — as well as finding them an agreeable habitat.

Lichens serve innumerable other purposes in nature. Birds use them to build nests. Golden plovers on St Lawrence Island in the Behring Sea off Alaska, for instance, fabricate their nests from the worm-like strands of *Thamnolia vermicularis*. British birds that employ lichens as constructional materials include species of hawfinch, goldcrest and long-tailed tit. And just as the combination of white *Thamnolia* with dark lichens and rocks helps to camouflage the golden plover's nest, many other creatures decorate their bodies with lichens to avoid predators. Beetles in New Guinea are especially resourceful, covering themselves in lichens and fungi to escape the attention of their enemies.

As well as finding them intrinsically interesting for the scientific study of symbiosis, man has exploited lichens

in innumerable practical ways. Several northern hemi-
sphere peoples cook and eat them — and for the Eskimoes
of northern Canada the partly digested lichens found in
the caribou stomach during winter are a prized delicacy.
Lichens also provide invaluable emergency rations. In
December 1972 a Canadian pilot lived on them, together
with glucose from an emergency kit, for twenty-three
days before being rescued after an air crash. One species,
Umbilicaria esculenta, is sold commercially as a food in
Japan, and during the last war Russia operated a factory
which used lichens to make sugar.

Folk medicines — laxatives, cough cures and countless
other nostrums — have long been derived from lichens.
Recently, however, drugs produced by them have been
investigated scientifically, and it begins to seem that they
have as yet been grossly under-exploited for their therapeutic versatility. One lichen antibiotic, usnic acid, has
been marketed in Germany, Finland and Russia for
treating such skin conditions as athlete's foot, ringworm
and lupus. It has also proved effective against fungus
diseases in plants. A remarkable feature of lichens as
sources of antibiotics is the handsome scale of their productivity. Some contain as much as five per cent by weight
of usnic acid.

Lichens provide raw materials for the perfume industry. They also yield many brown, purple and red dyes —
particularly in the Hebrides, where they are used in
making authentic Harris tweed. Litmus, the oldest chemical indicator, which turns red in acid and blue in alkali,
is still manufactured from a species of *Roccella.* And
lichens were formerly employed in the art of embalming.
But perhaps their most intriguing modern application
is in lichenology — the dating of rock surfaces by measuring colonies of lichens and thus estimating the time the
rocks have been exposed. The famous megaliths of Easter
Island have been studied using this technique. By comparing old photographs of the stones with present-day
measurements, it was possible to gauge the rate at which
lichens had been growing, and thus to estimate the date

when the megaliths were first built. The conclusion —
430 years ago — gave a younger age than had been expected, but it has been partially confirmed by Thor Heyer-
dahl's researches. Similar methods have been applied to
monitor the movements of glaciers, by observing lichens
growing in the moraines.

It would be difficult to imagine a group of creatures
with more diverse benevolence than the lichens. Yet,
slowly but inexorably, they are being allowed to disappear from the landscape. Time and time again, they
have been the casualties of industrialization and apathy.
At a public inquiry that preceded the building of a new
aluminium reduction plant near Holyhead in Anglesey
in 1968, the protesters — principally the British Lichen
Society and a small local residents' association — were
opposed by Rio Tinto-Zinc, the Kaiser Chemical Corporation and a consortium of other big companies. Against
the vast financial and legal and expert resources of the
consortium, there was no chance whatever of the opposi-
tion case — including the certain lethality of the factory
for lichens — being presented equally effectively. The
industrial consortium was given nine days to argue its
case, the objectors two days. No contest.

Extinction for smallpox?

But are there not virulent germs, unlike the friendly
lichens, for which we would welcome extinction? Louis
Pasteur dreamed of the possibility of wiping pathogenic
microbes from the face of the earth, and many medical
microbiologists and public health workers over the past
century have been inspired by that objective. Is it either
feasible or desirable? In general, clearly not. René Dubos
has pointed out that a totally aseptic world would be both
dangerous and dull. As we saw in the last chapter, a broad
perspective on infection and infectious disease makes
us at least hesitate in our indiscriminate anti-microbial
warfare, and compels us to re-examine our sanitary
enthusiasms.

What, though, of hardened microbial reprobates? What

of a villain like the smallpox virus? In 1967, the World
Health Organization embarked on a major crusade against
smallpox, a formidable killing disease and one of man-
kind's most feared scourges down the ages. The disease
was then endemic in thirty countries and was being
regularly imported into many more. Although it had been
eliminated from Europe and North America, chiefly by
mass vaccination, there were still about 2,500,000 cases
each year in Brazil and across vast areas of Asia and Africa.

Since that time, successive press releases from WHO
headquarters in Geneva have recorded the astonishing
achievements of the smallpox programme. Based on
vaccination and surveillance to detect and isolate any
cases that do occur, the crusade has been by far the most
triumphant of its sort in history. The last case of smallpox
in South America was reported in April 1971, and by
August 1973 a WHO Commission was able to pronounce
the total eradication of the disease from South America.
Indonesia's last case was recorded in January 1972, and in
April 1974 another WHO Commission announced that
that country too was now free of the disease. There was a
temporary setback in 1973 when the total number of cases
throughout the world doubled from the previous year's
65,000. This was due largely to the breakdown of public
health services as a result of disturbances in Bangladesh,
in northern India and in two of the four provinces of
Pakistan, where renewed epidemics occurred. But worldwide the number of countries with endemic smallpox had
declined to six by mid-1972, and by January 1974 they
were limited to Bangladesh, India, Pakistan and Ethiopia.
By November 1974 Pakistan appeared to be free of the
disease, and in a statement issued for World Health Day,
7 April 1975, the WHO announced: 'Victory is now in
sight in the World Health Organization's campaign to
vanquish smallpox from the earth. With India pronounced free in August and Bangladesh in November, only
Ethiopia remained. By 1978, WHO declared, smallpox
could probably be considered totally extinct.

The principal reasons why this achievement has been possible are the solid immunity that follows vaccination, which thus makes it possible to interrupt transmission of the disease, and the fact that the virus does not lurk in another animal, in the soil or elsewhere, in an infectious form. As with earlier successes in specific countries, the World Health Organization will doubtless wait several years before claiming that smallpox has indeed become totally extinct. Such caution is wise — but there is no doubt whatever that extinction is theoretically a perfectly sound possibility. What makes the prospect entirely unprecedented is that this will be the first time in history when man has been able to obliterate, for all time and by conscious, rational choice, a particular form of life: the smallpox virus. Many creatures have become extinct because of accident or shameful neglect, never before through calculated decision. Should the WHO be applauded for pioneering this new form of genocide? Or should the conservationists have moved in to call a halt?

Conserving pathogens

The very idea of protecting the smallpox virus or any other dangerous microbe does, of course, sound quite ludicrous — but only because of the minute size and innate nastiness of such forms of life. Some of us who would say farewell quite happily to a virulent virus or bacterium may well have qualms about eradicating forever a 'higher' animal — whether rat or bird or flea — that passes on such microbes to man. *Plasmodium vivax,* the protozoon responsible for malaria, might not be considered a great loss if it were to become extinct, but what of the mosquitoes that transmit malarial parasites to their human victims? Where, moving up the size and nastiness scale (rabies virus, typhoid fever bacilli, malarial parasites, bilharzia worms, locusts, rats...), does conservation become important? There is, in fact, no logical line that can be drawn. Every one of the arguments adduced by conservationists applies to the world of vermin and pathogenic microbes just as they apply to whales, gentians and flamingoes. Even the tiniest and most virulent virus qualifies.

Take the aesthetic argument. To an enthusiast, our most formidable foes in the microbial world, real and imagined, are every bit as attractive as butterflies and koala bears are to their protectors. A casual glance at their portraits taken under the electron microscope confirms this. The fact that they are tiny is irrelevant.

Next, while all species of life are valuable in research, none are more so than the microbes (see chapter 9). Most of the momentous discoveries in biology in recent years have come from investigations on bacteria and viruses, including pathogenic species. The loss of any one species could be an appalling loss not only for studies in comparative biochemistry and genetics but also for fundamental work on the origin and nature of life.

Then there are straightforward guilt feelings. Because man is the only product of evolution able to take conscious steps, whether logical or emotional, to influence its course, we have a responsibility to see that no other species is wiped out. But if we feel twinges of guilt about the impending extinction of large creatures, why should we feel differently about small ones? Conservationists lavish just as much time and energy on butterflies as they do on elephants. Why discount the microbes?

In many cases too, the ecological argument is a persuasive one. The dangers of 'upsetting the balance of nature' by deleting a single living component from a natural ecosystem are well known. While this risk may not apply to the smallpox virus (which, as far as we are yet aware, we can vanquish with impunity) there are many other occasional pathogens whose obliteration would have deleterious repercussions.

The most compelling argument of all, however, is that we may well need the smallpox virus some day, sooner or later — to help us in fighting disease and for other practical purposes. The incredible complexity of even such a miniscule form of life as the smallpox virus, particularly its DNA, could never be rebuilt in the laboratory. It represents the end-product of millions of years of organic evolution and is totally unique. Despite those occasional reports in the press about scientists creating life (which, even when they are accurate, are never more than stories about the reassembly or modification of existing living material),there is no possibility whatever of the fabrication from laboratory chemicals of a life-form remotely resembling any natural creature in its complexity.

Engineering with genes

The field in which even such pathogenic microbes as the smallpox virus could prove of unique value is the newly emerging science of 'genetic engineering'. Still hotly controversial because of its possible dangers and malevolent applications, genetic engineering covers a range of techniques for bringing together genes that would not otherwise combine in nature. DNA from different species, including that from animal and microbial cells, can be joined, extending enormously the potential manipulations of living processes in medicine, agriculture and microbiological industry. We stand only on the threshold of development of this new science, but its potential in fabricating novel genetic combinations for practical exploitation is enormous.

Nobel laureates Joshua Lederberg and Edward Tatum were the first to publicize the idea of genetic engineering, which was taken up and explored initially by Dr Stanfield Rogers, a geneticist at Oak Ridge National Laboratory, Tennessee, in the late 1960s. One of his earliest achievements was to add an extra artificial message to the nucleic acid in a plant virus and thus cause it to behave abnormally in infected plants. He and his colleague Dr Peter Pfuderer extracted RNA from tobacco mosaic virus and incubated it with chemicals they believed would alter the coding units in such a way that they would direct the synthesis of a different sort of protein from usual. The protein should contain excess lysine. So it proved. Leaves

from plants infected with the modified RNA contained
some unusual substances, strings of up to five lysine
molecules joined together. In other words, information
can be added to hereditary material in the test-tube, and
the virus used as a vector to transmit the desired message.
Dr Roger's experiment was relatively crude, but since
then work by Dr Har Gobind Khorana and his colleagues
at the Massachusetts Institute of Technology have shown
that entire genes, constructed like those in natural cells,
can be fabricated in the laboratory.

The most obvious potential application of the technique in medicine is that of using a virus, primed with
the necessary genetic information, to treat patients with
enzyme deficiency diseases. These are hereditary conditions in which the body lacks the capacity to produce a
necessary enzyme. If the missing genetic message could
be incorporated into such a patient's cells, the disease
might be cured. Indeed, one of Dr Rogers' early findings was that something similar had already happened —
unknown to the people whose genes had been augmented.
The starting point for the discovery was the observation
that rabbits infected with a cancer-causing virus, the
Shope papilloma virus, carry a characteristic enzyme
that breaks down arginine in the body. As well as producing tumours, the virus introduces into the rabbit cells a
gene directing synthesis of this enzyme. The virus doesn't
cause tumours in man. Dr Rogers wondered whether,
despite this fact and despite the absence of any normal
signs of infection, people handling the virus might have
been infected to the extent of having the tell-tale gene
incorporated into their cells. This is what happens when
the virus is added to isolated tissue in the laboratory: there
are no tumours, no visible effects of infection, but the key
enzyme nonetheless appears.

In theory, the easiest way of tracing the enzyme in the
human body would be to measure arginine in the bloodstream. Presence of the enzyme, and thus of the virus
gene, could be betrayed by abnormally low levels of argi-

nine. When Dr Rogers tested blood from twenty-two scientists who had worked with the virus, he found significantly less arginine than in twenty randomly selected individuals. Many of the scientists had had only brief contact with the virus — in one case twenty years previously. Dr Robert Shope, who discovered the virus and had inoculated it into himself without ill effect, also re-examined serum from a blood sample he had taken after infection and had preserved by freezing thirty years earlier. If contained extremely little arginine.

The conclusion seems inescapable. A harmless virus can become a genetic passenger in man and thereby alter the body's chemistry. Though the observations with Shope papilloma virus concern a 'natural' phenomenon, they do indicate one possible scenario for the artificial modification of genetic potential using viruses. As well as correcting missing enzyme activities, viruses could theoretically be given such tasks as restoring normal DNA repair machinery in the skin disease Xeroderma pigmentosum (p. 187).

A quite different avenue is the exploitation of bacteria as miniature factories to manufacture hormones such as insulin. If, as seems likely, the animal genes coding for such substances can be introduced into bacteria, the microbes may be able to synthesize materials that are totally foreign to them. Pituitary growth hormone, used in treating certain sorts of dwarfism, and interferon, a substance of potential therapeutic value in repelling virus infection, may also be made in this way. The opportunities for synthesizing such substances using micro-organisms, growing in vessels like those employed in making antibiotics and industrial chemicals, is one of the most exciting prospects the pharmaceutical industry has faced for many decades.

Another potentiality is that of isolating the genes responsible for nitrogen fixation in bacteria, and splicing them into the DNA of crop plants such as cereals. Similar agricultural prospects include the introduction of genes that increase the efficiency of photosynthesis, and of those that enhance the nutritive value of plants and plant products. Verging towards the realms of science fiction, but again theoretically possible, is the idea of creating hybrid cells combining the photosynthetic capacity of an alga with the high nutritional value of an animal cell such as liver. Meat and veg. in one cell?

A word of caution is necessary in assessing such prospects. Though the latent benefits of genetic engineering are abundantly clear, so too are the unpredictable dangers. It is possible, for example, that novel, infectious microorganisms might be fabricated (by either design or accident) that would be hazardous to human or animal health, to agricultural productivity or to vital natural processes. With these unpredictable dangers in mind, a committee of the US National Academy of Sciences, chaired by Professor Paul Berg, called in 1974 for at least a temporary moratorium on further experimentation in some types of genetic engineering until the risks had been fully investigated. In Britain a working party under Lord Ashby considered the matter and early in 1975 recommended that, subject to certain practical safeguards in the laboratory, the experiments should continue 'because of the great benefits to which they may lead'. The debate still continues. What nobody has challenged, however, is the immense practical promise of this new field of microbial exploitation.

Organisms on demand?

One of the most staggering revelations from early work on the genetic code was the incredible amount of information carried in the DNA helix. Dr K. Attwood of Columbia University, New York, has recently cast a mathematical eye on this cornucopia of detail to explore the theoretical possibilities for new forms of life. He began (as do so many biologists) with *E. coli.* So complex is this tiny microbe that the number of possible combinations of units in its hereditary material is $2^{6,000,000}$ — a very large figure indeed. Many of these combinations would contain incompatible genes or would be sheer nonsense. Even allowing for this wastage, however, the number remaining is so big that only a minute proportion of those combinations — and thus of the organisms they specify — could ever have existed on the earth.

Turning to larger creatures, the number of units in their DNA rises by up to one hundredfold. Though this is a far more modest increase than we might have expected in comparing higher with so-called primitive organisms, the proportion of possible beasts that has actually existed becomes even smaller. This means that the evolutionary process has not sorted through the majority of the possible combinations so that those most efficient in a particular environment have had the opportunity to exist and survive. The ablest possible organisms do not exist today and have never existed.

'Nature does not know best', Sir Peter Medawar said in his superb 1959 Reith lectures on the Future of Man. 'Genetical evolution, if we choose to look at it liverishly instead of with fatuous good humour, is a story of waste, makeshift, compromise, and blunder.' True — and Medawar gave many examples to illustrate his point — but it can also be argued that, though the raw materials of evolution were products of chaos and confusion, the stern and ruthless hand of natural selection must have brought high efficiency out of aeons of randomness and waste. There was, after all, time for virtually limitless experimentation as the laws of chance threw up every imaginable genetic makeup from the blind lottery of mutation and gene shuffling.

Not so, if we are to believe Dr Attwood, whose work suggests that the numerical prospects for new organisms far exceed the range of creatures on earth, past and present. This theoretical reasoning has encouraged another biologist, Professor James Danielli, director of the Center for Theoretical Biology at the State University of New York in Buffalo, in his practical research into tailor-making organisms that have not previously existed. Professor Danielli is one of those research workers who has been reported, notably in November 1970, as having 'created life in a test-tube'. What he had actually done was to reassemble an amoeba from the nucleus of one, the membrane of another and the cytoplasm of a third. Mis-reports notwithstanding, Danielli is actively interested in fabricating novel organisms for industrial and other processes, using genetic engineering and allied techniques. 'Most of the organisms which could exist to fulfil the special demands of civilisation do not exist now,' he argues, 'but can be brought into being by using various combinations of life synthesis techniques.' In contrast to the natural world, with its species barriers, 'life synthesis techniques make it possible to explore all possible combinations of genes which are viable'. The theoretical potentialities for bettering nature in this way are astronomical. In addition to nitrogen-fixing cereals and tailor-made sewage scavengers, other possibilities that suggest themselves are cellular computers, new organisms to grow in the oceans, biological desalinators and crop plants for arid zones. The genetic reserves of present-day microbes would be essential for such work.

The range of techniques, including those of genetic engineering, needed for these manipulations are far from fully developed. There are possible hazards that must be assessed before the more bizarre experiments of this sort can be carried out. But the hypothetical opportunities are compelling. Whether, in fabricating and exploiting such exotic superbeasts, we shall be able to apply enough of that other prized product of organic evolution, biological wisdom, is another matter altogether.

Conserving germs

What *is* clear is the importance of conserving existing microbial resources because of genetic potential that could at any moment prove of inestimable value for human welfare. In this context, the smallpox virus, the plague bacillus and the bacteria that cause VD are potentially

as important as are microbes that fix nitrogen, photosynthesize, digest cellulose and accomplish so many
other useful tasks which we have already discovered
to be to our benefit. We simply do not know when part
of the genetic material of such micro-organisms will
suddenly become of vital practical importance. It is,
indeed, notable that the one microbe already used, albeit
unconsciously, in genetic engineering is a pathogen
that causes tumours in rabbits. Quite possibly the smallpox virus, like that of Shope papilloma, may turn out
to have genetic abilities that are useful to man.

Early in 1974 Dr Stan Martin, a microbiologist at the
National Research Council of Canada, Ottawa, published
in the international scientific journal *Nature* a proposal
for the establishment of regional culture collections
of micro-organisms throughout the world, and particularly in developing countries. Calling on the United
Nations to sponsor such a project, he pointed out the
urgent need to conserve the genetic resources of microbes,
in view of man's total dependence on microbial activity
and the potential uses to which such reserves might be
put in the future. Surprisingly, although microbiologists have always been in the habit of preserving cultures
of their organisms in a viable condition, there has been
nothing like the same interest in maintaining microbes
as a genetic resource as there has, for example, with cereal
varieties. That need is now amply clear. Hard necessity,
not sentiment as Edward Frankland suggested (p. 170),
is the justification for cherishing and conserving the
diversity of microbes.

Disappearing pathogens are automatic candidates
for consideration. Smallpox virus, agricultural pests
and other 'nuisance species', if and when they are eradicated from nature, will have to be confined in special
facilities (and displayed in micro-zoos where holiday
crowds can gawk at smallpox and the black death?). It
may be some time before other microbes go the way of
smallpox, but many once ubiquitous organisms, such

as the diphtheria bacillus, have been largely banished
from most Western countries. Measles has been eradicated
in several states in the US. In August 1974 the WHO
announced that even plague, and thus its causative bacterium *Pasteurella pestis,* was disappearing rapidly from the
world scene. Three months later, the *Lancet* lamented
in an editorial that 'the specialty of infectious diseases
has reached a low point', and at about the same time
Dr Edwin Kilbourne, of Mount Sinai School of Medicine,
New York, announced : 'It is time to capitalise on the
legacy of modern molecular biology in the deliberate
design and choice of the viruses with which we shall live
and which shall defend us.' For a hard-nosed medical
virologist to talk in such terms is indeed significant.
It is to be hoped that, in making the type of decisions
he mentions, we will show ecological sensitivity and
discernment.

Microbial tasks, predictable and hazy
Our future need for microbial aid is clear in many areas.
Food production is the paramount example. Like thirsty
shipwrecked mariners, afloat in an open boat without
fresh water and unable to exploit the abundance of the
oceans, we live surrounded by glucose, bound up in the inaccessible form of plant cellulose, and bathed in nitrogen:
neither can be used without assistance from micro-organisms. In exploiting both resources, by stratagems conventional and heterodox, we shall be increasingly dependent on microbial munificence in future. The microbes
will be our trusty allies in many other fields touched
on in this book — though the range of talents on offer
is so diverse as to make it difficult to predict where the
greatest emphasis will lie. In some cases, like the acetone-
butanol and alcohol fermentations (chapter 8), the technology already exists, and we only have to apply it to the
full. Elsewhere, recently discovered possibilities await
exploitation.

Conservation of the genetic potential of micro-organisms

is vital to these developments, as it is to the unforeseeable tasks of which we may find microbes willing exponents in the future. Their latent skills can perhaps be guessed at from the range of their present accomplishments: mining for metals, ridding coal mines of methane, synthesizing vitamin B_{12}, cleaning up oil spills, revealing the secrets of life and serving as tools of medical research, creating perfumes, leavening bread, assaying chemicals, making oral contraceptives, monitoring air pollution, controlling insect pests, dating Easter Island megaliths, destroying insecticides in the soil, generating methane for heating, helping to manufacture soft-centred chocolates, retting flax, ensuring the proper development of the heart and intestines, and thousands more.

New possibilities are continuously being suggested. Three recent ideas are to employ bioluminescent bacteria to detect heroin (proposed by the New York Police Department), to use bacteria to remove waste urea in victims of kidney failure (proposed by Dr Kai Setälä of the University of Helsinki), and to seed the planet Venus with algae, rendering it a pleasant place to live (proposed by Carl Sagan). The police officers' bacteria would glow in the presence of heroin. Dr Setälä's scavengers would operate in the gut, extracting the noxious wastes accumulating in the bloodstream. Professor Sagan's algae would photosynethsize and convert some of Venus's abundant carbon dioxide into oxygen, causing the planet to cool from its present sizzling temperature and triggering what Poul Anderson has called 'The Big Rain', making the planet more like Earth.

Whatever the feasibility of such notions, they do serve to indicate areas of untapped and even unforeseen microbial potential. The schemes of man aside, we are still becoming aware of the manifold functions and symbioses of microbes in the natural world — like the vital role of ethylene-producing fungi in regulating the soil flora, which was discovered only during 1974, and the emerging knowledge of our indebtedness to the lush population

of micro-organisms inhabiting the human body. The
vividly clear lesson is that we still have much to learn
about the microbial under-pinning of the natural world
and of human existence, and that we should cherish the
microbes for their present and future potential in furthering our welfare.

The real priorities

All of which renders the more starkly absurd the closing
down, in December 1958, of Britain's only and outstandingly important national laboratory devoted to economic
and industrial microbiology. The laboratory, at Teddington, originated with a recognition, just after the war, of
Britain's backwardness in industrial microbiology — one
reason why penicillin, though discovered in Britain, had
to be developed commercially in the United States. In
later years the laboratory turned its attention to such
projects as using bacteria to produce hydrogen sulphide,
and thus sulphur, from sewage. Despite developing a
large-scale process of this sort, and pursuing excellent
pioneering work in effluent disposal and methane generation, despite a national outcry from biologists and despite
the future economic potential of such studies, the laboratory was abolished by the then Department of Scientific
and Industrial Research. As its director, Dr Kenneth
Butlin, recorded, protests were 'met with a silent, rock-
like obstinacy impervious to common sense'. And when,
in 1962, Dr Butlin proposed the setting up of a National
Laboratory of Economic Microbiology, his views were
again ignored by the administrators. In the new world of
the 1970s, in which microbiological assistance commends
itself with unprecedented force in coping with many of
our most pressing economic problems — and when, indeed, the very work pioneered at Teddington is being
urgently developed once more — the short-sightedness
of those days appears positively criminal.

But Britain is not unique in getting its priorities wrong.
Addressing the Third Conference on Global Impacts of

Applied Microbiology, in Bombay in December 1968, Professor Carl-Göran Hedén presented a number of options for harnessing microbiology for human welfare, and concluded with the following questions:

> Where do we find the best facilities for large-scale production of the pathogens which attack the insects and pests which destroy our biological resources? Who are the experts on the aerosol delivery of such agents and on the techniques available for stabilising them? Where are the geneticists who can make them resistant to the insecticides and other chemicals that might be employed in 'integrated control' programmes? Who have the facilities and the competence for producing large quantities of veterinary medicines? Who have experience of using attenuated strains for aerosol immunisation and know the logistics of mass vaccinations in general? Where can we learn about rapid diagnostic procedures suitable for field work?

The answer, of course, was that by far the biggest concentrations of applied microbiologists in the world at that time were to be found in government laboratories concerned with biological warfare. In the US some 3,000 people, including 430 medical doctors, were working on biological and chemical warfare at Fort Detrick. Another 1,700 at Pine Bluff Arsenal, Arkansas (using equipment valued in 1966 at 138 million dollars) were producing the country's CBW munitions, which were field-tested at the Dugway proving grounds in Utah, where another 1,600 personnel worked. The Soviet Union, Professor Hedén reasonably suggested, probably had comparable numbers involved in the same type of work. By contrast, the World Health Organization at that time commanded a total staff of just 3,500.

And today? It is impossible to be sure of one's facts in such an evil and secret field, but the signs are bad. In 1969 President Nixon renounced both biological and chemical warfare, but the Department of Defense continued to

finance a multi-million dollar research programme on
infectious disease. Germ-producing facilities were not
dismantled. In October 1971 Nixon revoked biological
warfare again and said that Fort Detrick would be turned
over to cancer research, yet the laboratory went on with
its biological warfare research, which it continued to call
'defensive'. The giant germ-breeding plant at Pine Bluff
has only been sealed off, not dismantled. Even when the
US ratified the Geneva Protocol early in 1975, there was
no concrete evidence that the country's CBW capability
had been destroyed. In Britain too, though there were
rumours that the Microbiological Research Establishment at Porton Down might be turned over to entirely
'civil' research after the great powers signed the Convention on Biological Weapons early in 1972, little more has
been heard of that idea. Instead of passing to the Department of Health, as had been hinted at the time, Porton
remains a Ministry of Defence establishment.

'Ours is a busy, vexed, quarrelsome world', were the
words with which Dr Hugh Nicol began his Penguin
Special *Microbes by the Million,* published in 1939 and still
the most lucid popular account of the beneficence of microbes in the soil. They are uncommonly apposite as a
commentary on what man has done to harness the abundant promise of the microbes. We have become weary
of comparing expenditure on nuclear armaments with
that on the many urgent and harrowing needs of the
world. But to note that the most lavishly funded mission-
oriented projects ever mounted in applied microbiology
have also been devoted to warfare rather than welfare
brings one close to despair. Cynical expectation that
nothing will ever change is no longer an option. In a
world where the most fundamental, inescapable and
universal long-term enemies are clearly not those of petty
nationalism, we must believe that we can change the
priorities.

Bibliography

The following list contains research papers and books I have consulted (including those specifically cited in the text) together with others that may be of further interest to the reader.

Alexander, M., 'Biodegradation: problems of molecular recalcitrance and microbial infallibility', *Advances in Applied Microbiology*, 7 (1965), p. 35.

Anderson, E. S., 'The truth about typhoid', *The Times* 31 December 1971.

Avery, O. T., *et al.*, 'Studies on the chemical nature of the substances inducing transformation of pneumococcal types', *Journal of Experimental Medicine*, 79 (1944), p. 137.

Barghoorn, E. S., and Tyler, S. A., 'Micro-organisms from the gun flint chert', *Science*, 147 (1965), p. 563.

Berman, G. A., and Murashige, K. H. (eds), *Synthetic Carbohydrate, an aid to nutrition in the future*, NASA/ASEE, 1973.

Birch, G. G., *et al.* (eds) *Food from Waste* Applied Scientific Publishers, 1976.

Bisset, K., 'Most important invention', *New Scientist*, 13 (1969), p. 426.

Bone, Q., and Home, N., 'Lessons from the *Torrey Canyon*', *New Scientist*, 39 (1968), p. 492.

Breznak, J. A., *et al.*, 'Nitrogen fixation by termites', *Nature*, 244 (1973), p. 577.

Brock, T. D., *Biology of Micro-organisms* Prentice-Hall, 1970. 'Limits of microbial existence: temperature and pH', *Science*, 169 (1970), p. 1316.

Brown, B. S., *et al.*, 'Chemical and biological degradation of waste plastics', *Nature*, 250 (1974), p. 161.

Brown, M. E., 'Plant growth substances produced by micro-organisms of soil and rhizosphere', *Journal of Applied Bacteriology*, 35 (1972), p. 443.

Brown, M. E., and Jackson, R. M., 'Bacterial inoculation of non-leguminous crops', *NAAS Quarterly Review*, no. 70 (1965), p. 69.

Bryson, V. (ed.), *Microbiology yesterday and today* Rutgers Institute of Microbiology, 1959.

Bulloch, W., *The history of bacteriology* Oxford University Press, 1938, reprinted 1960.

Burges, H. D., and Hussey, N. W., *Microbiological control of insects and mites* Academic Press, 1971.

Burnet, F. M., *Natural history of infectious disease* Cambridge University Press, 1972, 4th edn.

Butlin, K., 'Prospects in industrial microbiology', *New Scientist*, 14 (1962), p. 804.

Campbell, J., and Lengyel, J., 'Evolution of a second gene for β-galactosidase in *Escherichia coli*, *Proceedings of the National Academy of Sciences*, 70 (1973), p. 1841.

Carlson, P. S., and Chaleff, R. S., 'Forced associations between higher plants and bacterial cells *in vitro*', *Nature*, 252 (1974), p. 393.

Carson, R., *Silent Spring* Houghton Mifflin, 1962, Hamish Hamilton, 1963.

Chang, T. M. S., *Artificial cells* Charles C. Thomas, 1972.

Chapman, V. J., and D. J., *The Algae* Macmillan, 1973, 2nd edn.

Charney, W., 'Microbes as chemical reagents', *New Scientist*, 43 (1969), p. 10 suppl.

Child, J. J., 'Nitrogen fixation by a *Rhizobium* sp. in association with non-leguminous plant cell cultures', *Nature*, 253 (1975), p. 350.

Clarke, R., *We all fall down — the prospect of biological and chemical warfare* Allen Lane, 1968.

Coates, M. E., 'Nutrition and the microflora of the gastro-intestinal tract', *British Nutrition Foundation Bulletin*, no. 9, (1973), p. 34.

Conant, J. B., *The chemistry of organic compounds* Macmillan, 1934.

Courage, R. H., 'A mystery and a science', *The Times* 22 April 1968.

Dagley, Stanley, in *Degradation of synthetic organic molecules in the biosphere* National Academy of Sciences, 1972, p. 338.

Danielli, J. F., 'Artificial synthesis of new life forms', *Bulletin of the Atomic Scientists*, 28 (1972), p. 20.

Davis, P. (ed.), *Single cell protein* Academic Press, 1974.

Dixon, B., *What is science for?* Collins, 1973, Harper & Row, 1974, Penguin, 1976.

Dobell, C., *Antony van Leeuwenhoek and his 'little animals'* Dover Publications, 1960.

Dubos, R., *Mirage of health* Allen & Unwin, 1960, Harper & Row, 1971.

Duddington, C. L., *Micro-organisms as allies* Faber, 1961.

Fineberg, R., 'United States continues work on CBW', *New Scientist*, 56 (1972), p. 501.

Fogarty, W. M., and Ward, O. P., in D. J. D. Hockenhull (ed.), *Progress in Industrial Microbiology*, vol. 13 (1974), p. 59.

Fogg, G. E., *et al.*, *The blue-green Algae* Academic Press, 1973.

Frankland, E., 'On chemical changes in their relation to micro-organisms', *Journal of the Chemical Society*, 47 (1885), p. 159.

Frobisher, M., *Fundamentals of microbiology* W. B. Saunders, 1968, 8th edn.

Gause, G. F., *Microbial models of cancer cells* North-Holland, 1966.

Gilbert, O. L., 'New tasks for lowly plants', *New Scientist*, 46 (1970), p. 288.

Gray, T. R. G., and Williams, S. T., *Soil micro-organisms* Oliver & Boyd, 1971.

Gregory, P. H., *Microbiology of the atmosphere* Leonard Hill, 1973, 2nd edn., Halsted Press, 1973.

Grigg, H., 'Bacteria for nickel extraction', *Search*, 5 (1974), p. 270.

Hale, M. E., *The biology of lichens* Edward Arnold, 1970, American Elsevier, 1970.

Harada, T., 'The role of micro-organisms in food production', *Impact of Science on Society*, 24 (1974), p. 171.

Hare, R., *The birth of penicillin* Allen & Unwin, 1970.

Hedén, C-G., 'Applied microbiology — for life or for death',

Third Conference on Global Impacts of Applied Microbiology, Bombay, 1969.

Socio-economic and Ethical Implications of Enzyme Engineering
(IFIAS Study no. 1) Nobel House, Stockholm, 1974.

Hegner, R., *Big fleas have little fleas or Who's who among the
Protozoa* Dover Publications, 1968.

Hobson, P. N., 'Digesting the indigestible', *New Scientist*, 40
(1968), p. 142.

Hudson, B. J. F., 'New protein foods in the UK', *Chemistry
and Industry* (18 March 1972), p. 251.

Hughes, D. E., 'Towards a recycling society', *New Scientist*,
61 (1974), p. 58.

Hughes, D. E., and Rose, A. H. (eds.), *Microbes and biological
productivity* Cambridge University Press, 1971.

ICI, *New protein* Kynoch Press, 1974.

Jefferys, E. G., 'The Gibberellin fermentation', *Advances in
Applied Microbiology*, 13 (1970), p. 283.

Jenkinson, D. S., 'Studies on the decomposition of plant
material in soil', *Journal of Soil Science*, 16 (1965),
p. 104.

Jeon, K. W., 'Development of cellular dependence on infective organisms: micrurgical studies in amoebas',
Science, 176 (1972), p. 1122.

Jones, K., and Thomas, J. G., 'Nitrogen fixation by the rumen
contents of sheep', *Journal of General Microbiology*, 85 (1974),
p. 97.

Kilbourne, E. D., 'Virus research', *Science* 184 (1974), p. 414.

Le Roux, N. W., 'Mining with microbes', *New Scientist*, 43
(1969), p. 12 suppl.

'Mineral attack by microbiological processes', in
J. D. A. Miller (ed.) *Microbial Aspects of Metallurgy* Medical
and Technical Publishing, 1971.

Le Roux, N. W., *et al.*, 'Bacterial oxidation of pyrite', *Tenth
International Mining Processing Congress* Institution of Mining and Metallurgy, 1973.

Lechevalier, H. A., and Solotorovsky, M., *Three centuries of
microbiology* McGraw Hill, 1965.

Lewin, B., 'New genes for old?' *New Scientist,* 54 (1972), p. 122.

Lloyd George, D., *War Memoirs* Odhams, 1933-34.

Margalith, P., and Schwartz, Y., 'Flavor and micro-organisms', *Advances in Applied Microbiology,* 12 (1970), p. 36.

Martin, S. M., 'A case for regional culture collections of micro-organisms', *Nature,* 247 (1974), p. 431.

Marx, J. L., 'Insect Control (II): Hormones and viruses', *Science,* 181 (1973), p. 833.

Meadow, P., and Pirt, S. J. (eds), *Microbial growth* Cambridge University Press, 1969.

Medawar, P. B., *The future of Man* Methuen, 1960.

Miall, L. M., 'Microbes as acid manufacturers', *New Scientist,* 43 (1969), p. 8 suppl.

Michaelis, A. R., *The Weizmann centenary* Anglo-Israel Association, 1974.

Moseley, B. E. B., 'Keeping DNA in good repair', *New Scientist,* 41 (1969) p. 626.

Nicol, H., *Microbes by the million* Penguin, 1939.
Microbes and us Penguin, 1955.

Oswald, W. J., 'Solar energy fixation with algal-bacterial systems', *Compost Science,* 15 (1974), p. 20.

Perutz, M. F., in *Encyclopaedia of Life Sciences,* vol 1 Doubleday 1965.

Peters, J. A. (ed.), *Classic papers in genetics* Prentice-Hall, 1959.

Porteous, A., 'Sweet solution to domestic refuse', *New Scientist,* 50 (1971), p. 736.

Postgate, J., *Microbes and Man* Penguin, 1969.

Potter, M. C., 'Electrical effects accompanying the decomposition of organic compounds', *Proceedings of the University of Durham Philosophical Society,* 4 (1911), p. 260.

Powell, A. J., and Bu'Lock, J. D., *Projects and prospects in industrial fermentation,* Octagon Papers no. 1, University of Manchester, 1974.

Rainbow, C., and Rose, A. H., *Biochemistry of industrial organisms* Academic Press, 1965.

Report of the Working Party on the Experimental Manipulation of the Genetic Composition of Microorganisms Cmnd 5880, HMSO, 1975.

Richardson, D., *The vanishing lichens* David & Charles, 1975.

Rogers, S., 'Skills for genetic engineers', *New Scientist*, 45 (1970), p. 194.

Rose, A. H., *Industrial microbiology* Butterworth, 1970, 2nd edn.

Rose, A. H., and Harrison, J. S., *The yeasts*, vol. 3 *Yeast technology*, Academic Press, 1970.

Rose, S. (ed.), *Chemical and biological warfare* Harrap, 1968, Beacon Press, 1969.

Rosebury, T., *Life on Man* Secker & Warburg, 1969, Viking Press, 1969.

Round, F. E., *The biology of Algae* Edward Arnold, 1973, 2nd edn.

Sagan, C., 'The planet Venus', *Science*, 133 (1961), p. 24.

Scholtissek, C., and Rott, R., 'Effect of temperature on the multiplication of an influenza virus', *Journal of General Virology*, 5 (1969), p. 283.

Scowcroft, W. R., and Gibson, A. H., 'Nitrogen fixation by *Rhizobium* associated with tobacco and cowpea cell cultures', *Nature*, 253 (1975), p. 351.

Setala, K., *et al.*, 'Uraemic waste recovery II: *In vitro* studies', in J. Stewart Cameron (ed.), *Dialysis and Renal Transplantation* Pitman Medical, 1973.

Sherwood, M., 'Single-cell protein comes of age', *New Scientist*, 64 (1974), p. 634.

Sisler, F. D., 'Biochemical fuel cells', in D. J. D. Hockenhull (ed.), *Progress in Industrial Microbiology*, vol. 9 (1971), p. 1.

Sleigh, M., *The biology of Protozoa* Edward Arnold, 1973.

Smith, A., *The body* Allen & Unwin, 1968, Avon, 1969.

Smith, A. M., and Cook, R. J., 'Implications of ethylene production by bacteria for biological balance of soil', *Nature*, 252 (1974), p. 703.

Smyers, W., 'Gasoline substitutes', *Science,* 183 (1974), p. 698.

Stanier, R. Y., *et al.*, *General Microbiology* Macmillan, 1971, 3rd edn.

Starkey, R. L., 'Microbiology and the microbiologist', *Bacteriological Reviews*, 27 (1963), p. 243.

Stewart, G. T., 'Limitations of the germ theory', *Lancet*, 2 (1968), p. 1077.

Tinker, J., 'Bugs in the pipes', *New Scientist*, 49 (1971), p. 429.

Vallery-Radot, R., *The life of Pasteur* Dover Publications, 1960.

Wade, N., 'Insect viruses: a new class of pesticides', *Science*, 181 (1973), p. 925.

Waid, J. S., 'The possible importance of transfer factors in the bacterial degradation of herbicides in natural ecosystems' *Residue Reviews*, 44 (1973), p. 65.

Waksman, S., *My life with the microbes* Simon & Schuster, 1954, Robert Hale, 1958.

Walker, N. (ed.), *Soil Microbiology* Butterworth, 1975.

Walsh, J. H., 'The plastics disposal problem', *Biologist*, 19 (1972), p. 141.

Watson, J. D., and Crick, F. H. C., 'Molecular structure of nucleic acids', *Nature*, 171 (1953), p. 737.

Wingard, L., 'Enzyme engineering — a global approach', *New Scientist*, 64 (1974), p. 565.

World Health Organization, *WHO Expert Committee on Small-pox Eradication* Technical Report Series no. 493, 1972.
The Use of Viruses for the Control of Insect Pests and Disease Vectors, Technical Report Series no. 531, 1973.
'Progress in smallpox eradication', *WHO Chronicle* 28 (1974), p. 359.
Smallpox, point of no return 1975.

Zaborsky, O., 'Nine lives for trapped enzymes', *New Scientist*, 57 (1973), p. 719.

Zinder, N. D., and Lederberg, J., 'Genetic exchange in *Salmonella*', *Journal of Bacteriology*, 64 (1952), p. 679.

Zinsser, S., *Rats, lice, and history* Little, Brown, 1935, Routledge, 1942, 4th edn.

Index